Community-Supported Agriculture in Central and Eastern Europe

This book analyses Community-Supported Agriculture (CSA) in Central and Eastern Europe, providing a meaningful contribution to the broader discourse on sustainable agriculture and rural development in post-communist states.

Drawing on comparative insights between Western and Central-Eastern European CSA models, the book addresses both structural and institutional factors influencing the functioning of CSA and offers a comparative perspective on the implementation of this model in Western and Central-Eastern Europe. Particular attention is given to the challenges and opportunities for its development, considering the region's socio-economic characteristics, institutional constraints, the level of social capital, and the growing consumer interest in local and organic food. The book draws on a detailed empirical study conducted in Poland, complemented with comparative insights from countries such as Hungary, the Czech Republic, and Romania, to identify common features and barriers to CSA development in post-communist settings. As the largest post-communist country in the European Union, Poland offers a valuable lens through which broader regional dynamics can be analyzed. The book explores the unique trajectories and structural challenges faced by CSA initiatives in transitional economies, identifying critical barriers and emerging opportunities in the development of collaborative food systems, highlighting the roles of both producers and consumers. It bridges the supply and demand dimensions of alternative food networks, offering a multifaceted interpretation of CSA's potential in reshaping agrifood relations in the region.

By combining conceptual analysis with practical policy relevance, this book will be of interest to students and scholars of food and agricultural, rural development and sociology and environmental economics. It also provides actionable insights for policymakers engaged in designing resilient local food systems and promoting rural innovation.

Magdalena Raftowicz is an Associate Professor at the Institute of Spatial Management at the Wrocław University of Environmental and Life Sciences, Poland.

Mirosław Struś is an Assistant Professor at the Institute of Economic Sciences at the University of Wrocław, Poland.

Earthscan Food and Agriculture

Pathways to African Food Security
Challenges, Threats and Opportunities towards 2050
Edited by Michiel de Haas and Ken Giller

Multiple Capitals Approach for Upland Agricultural Systems
Lois Mansfield

Agricultural Land Use
Structural Transformations, Environment Challenges, Planning and Policy
Edited by Jerzy Bański

Sustainable Production of Root and Tuber Crops
Edited by Ravinder Kumar, Milan Kumar Lal, Rahul Kumar Tiwari and Brajesh Singh

Agrifood Regimes in Greece
Sociotechnical Dynamics and Sustainable Transitions
Stathis Arapostathis, Sotiris Alexakis, Vasiliki Karantzavelou and Yannis Fotopoulos

Nourishing Scotland
Pete Ritchie

Community-Supported Agriculture in Central and Eastern Europe
Transitional Economies in Post-Communist Europe
Magdalena Raftowicz and Mirosław Struś

For more information about this series, please visit: www.routledge.com/Earthscan-Food-and-Agriculture/book-series/ECEFA

Community-Supported Agriculture in Central and Eastern Europe

Transitional Economies in Post-Communist Europe

Magdalena Raftowicz and Mirosław Struś

Routledge
Taylor & Francis Group
LONDON AND NEW YORK

earthscan
from Routledge

First published 2026
by Routledge
4 Park Square, Milton Park, Abingdon, Oxon OX14 4RN

and by Routledge
605 Third Avenue, New York, NY 10158

Routledge is an imprint of the Taylor & Francis Group, an informa business

British Library Cataloguing-in-Publication Data
A catalogue record for this book is available from the British Library

ISBN: 978-1-041-25120-0 (hbk)
ISBN: 978-1-041-25121-7 (pbk)
ISBN: 978-1-003-74519-8 (ebk)

DOI: 10.4324/9781003745198

Typeset in Times New Roman
by SPi Technologies India Pvt Ltd (Straive)

Contents

Introduction

Community-Supported Agriculture (CSA) is an innovative model of cooperation between producers and consumers that, over the past decades, has become a significant component of alternative food systems in many countries worldwide. Originating in Japan and the United States, this concept has developed dynamically in Western Europe, where it has found fertile ground among advocates of sustainable agriculture, short supply chains, and local, organic food production. In Central and Eastern Europe, the model remains in its early stages of development, raising questions about its potential and the barriers to adaptation within the context of post-communist socio-economic conditions.

The main aim of this monograph is to explore whether the concept of CSA is reflected in the realities of Central and Eastern European countries, and whether there is a need to develop a new CSA model more adequately suited to local socio-economic contexts.

The analysis addresses both structural and institutional factors influencing the functioning of CSA and offers a comparative perspective on the implementation of this model in Western and Central-Eastern Europe. Particular attention is given to the challenges and opportunities for its development, taking into account the region's socio-economic characteristics, institutional constraints, the level of social capital, and the growing consumer interest in local and organic food. In this context, Poland represents a highly relevant case for detailed examination.

The empirical study was conducted in Poland—a country where agriculture and the entire agri-food sector continue to play a significant role in the national economy. In terms of agricultural land area, Poland ranks third within the European Union and is simultaneously the largest agricultural producer among the newly acceded member states. At the same time, Poland is characterized by a highly fragmented farm structure (despite ongoing processes of land concentration), which theoretically predisposes it to the development of alternative forms of agricultural production compared to industrial (conventional) farming—particularly CSA.

DOI: 10.4324/9781003745198-1

The fact that the empirical study was carried out in Poland implies that the proposed solutions aimed at improving the functioning of CSA and enhancing its broader importance are most closely aligned with Poland's socio-economic and institutional conditions. Nevertheless, through an extensive review of the literature, the authors also incorporated into the developed model the specific characteristics of the Central and Eastern European region, arising, among other factors, from shared historical experiences—most notably the dual systemic transformation: first, after 1945, from a market economy to a centrally planned one, and subsequently, after 1989, from a centrally planned to a market economy.

This research draws upon a broad range of methodological approaches, including:

- a critical review of the literature,
- a comparative analysis of CSA models in Western and Central-Eastern Europe,
- an analysis of statistical data, reports, strategic documents, and legal and policy frameworks,
- surveys of farmers participating in CSA,
- consumer surveys,
- in-depth interviews with agricultural support institutions,
- and an in-depth interview with a representative of a non-governmental organization.

Based on the research conducted thus far, the monograph poses three open research questions:

1. To what extent does the functioning of CSA in Central and Eastern Europe differ from that in Western Europe as a result of distinct historical, economic, and social conditions?
2. Why is it particularly difficult to build long-term relationships between producers and consumers in Central and Eastern Europe?
3. Does CSA in Central and Eastern European countries have the potential for further development, and what factors are likely to shape its future?

The monograph consists of five chapters, presenting Community-Supported Agriculture from theoretical, historical, and empirical perspectives:

Chapter 1 outlines the origins, conceptual framework, and functioning of CSA. It analyses the core principles of CSA and its global development, with particular emphasis on Western Europe. The model is compared with traditional food production and distribution systems, highlighting its role within the broader context of alternative agriculture.

Chapter 2 discusses the theoretical foundations of CSA in relation to selected rural development concepts. It introduces frameworks such as the neoliberal development paradigm, sustainable development theory, social capital theory, and innovation theory. These perspectives help elucidate the mechanisms underlying CSA and its potential influence on agricultural development.

Chapter 3 examines the potential and barriers to CSA development in Poland. It provides historical background and assesses the impact of post-1989 economic transitions on agriculture, along with key factors shaping CSA's growth. Special attention is paid to institutional barriers, levels of social capital, and market conditions.

Chapter 4 presents the results of empirical research conducted in Poland. It explores the perspectives of both farmers and consumers, focusing on their motivations, perceived benefits, and challenges associated with participating in CSA. The chapter also evaluates the role of institutions supporting agricultural development and proposes potential mechanisms to strengthen CSA support.

Chapter 5 offers recommendations for the future development of CSA in Central and Eastern Europe. It suggests possible policy directions, institutional strategies, and approaches to increasing public engagement with the model. In addition, it presents forecasts for CSA's future in light of ongoing socio-economic and political changes.

Community-Supported Agriculture represents one of the most significant alternatives to the dominant model of industrial agriculture. Its development in Central and Eastern Europe presents both opportunities and challenges, shaped by the region's unique historical, social, and institutional conditions.

It is our hope that the analyses, research findings, and recommendations presented in this monograph will contribute to a deeper understanding of the potential of Community-Supported Agriculture and its role in transforming food systems toward greater sustainability. We also hope this work will serve as a foundation for further research and policy initiatives supporting the development of this model in Central and Eastern European countries.

1 The Origins, Definition, and Operational Framework of Community-Supported Agriculture

1.1 The Origins and Concept of CSA

Community-Supported Agriculture (CSA) has its origins in Japan, where the *Teikei* system emerged in the 1960s (Kondoh, 2015). It was developed in response to growing concerns among Japanese consumers regarding food safety, particularly in the wake of mercury contamination resulting from the environmental disaster in Minamata (Gallaud and Laperche, 2016). The *Teikei* system was based on direct cooperation between farmers and consumers, bypassing intermediaries to ensure transparency and high food quality (Farnsworth et al., 1996). However, similar initiatives in other parts of Asia have remained relatively uncommon. In China, the first CSA farm was not established until 2009, although by 2016 the number had already risen to 254 (Tang et al., 2019; Wang et al., 2025).

From Japan, the model spread to Europe in the late 1970s, initially reaching Switzerland under the name *Les Jardins de Cocagne* and subsequently expanding to France, Germany, and Italy, where it achieved notable success (Samoggia et al., 2019). In the 1980s, the idea of CSA reached the United States through the efforts of German farmer T. Groh, who introduced a community-based subscription model. In this arrangement, consumers commit to financially supporting a farmer in return for regular deliveries of fresh produce (Groh and McFadden, 1990). It was during this period that the term "Community-Supported Agriculture" entered public discourse for the first time, reflecting the concept of direct community engagement in the financial and logistical support of local farms.

The term was intended to reflect the essence of the system: agriculture that is not only financially sustained by the community but also fosters a meaningful relationship between consumers and farmers, providing both parties with increased stability and food security. It highlights the reciprocal nature of the relationship among market participants, who share both the risks and benefits associated with agricultural production.

By 1986, two CSA groups were operating in the United States: Indian Line Farm in Massachusetts and Temple-Wilton Community Farm in

DOI: 10.4324/9781003745198-2

New Hampshire (Allen et al., 2006). In the decades that followed, the CSA model expanded rapidly—from 1,144 farms in 2005 (Adam, 2006) to as many as 12,000 in 2012, which accounted for approximately 0.6% of all farms in the United States (Vassalos et al., 2017). This growth was fueled by increasing consumer interest in organic food, sustainable agriculture, and support for local economies.

According to Schnell (2007), the early CSA model in the United States was strongly rooted in the principles of co-responsibility, solidarity, and trust between farmers and consumers. It offered an alternative to the globalized food system, enabling the restoration of ties to the place where food originates and bringing back the "farmer's face" to the production process. Over the past decade, however, this model has undergone significant evolution—becoming increasingly market-oriented and moving away from its original principles of partnership and community. Farms have progressively abandoned formal organic certification in favor of participatory guarantee systems or cooperation based solely on trust. Expectations regarding member retention and product diversification have also grown, while competition and pressure to deliver short-term value to consumers have intensified. The duration of partnerships has likewise diversified: alongside traditional seasonal arrangements, long-term contracts have emerged, offering greater financial stability and more durable relationships with members.

As noted by Woods and Tropp (2015) and Woods et al. (2017), many CSA farms have begun to operate like small enterprises, employing marketing and sales strategies characteristic of the conventional agri-food market. Similar observations are presented by Galt et al. (2017), who demonstrate that income-related factors and consumer expectations increasingly shape participation in CSA. Changing market conditions have led producer–consumer relations to take on a more service-oriented character—shifting the emphasis from solidarity toward customer satisfaction and loyalty, resulting in the gradual commercialization and weakening of the movement's communal dimension.

Since its inception, the CSA model has undergone significant evolution, incorporating various forms of cooperation—from collective on-farm labor and pick-up at distribution points to direct home deliveries.

As a result, a variety of CSA models has emerged, allowing researchers to classify them according to different analytical criteria. For example, Blättel-Mink et al. (2017) distinguish CSAs based on their underlying ideologies, categorizing them as "part of socio-political change," "a spiritual–community practice," or "a pragmatic–economic strategy." Likewise, Bîrhală and Möllers (2014) differentiate between "subscription-based CSAs" and "shareholder CSAs," depending on whether the initiative originated from producers or from consumers. Gruber (2020) further classifies CSAs as self-organized, participatory, or service-oriented, reflecting the diverse motivations that lead members to engage in such arrangements. Ultimately, however, each CSA is shaped by its specific context, the attitudes and preferences of its participants, its legal structure, the resources available, and the ways in which these elements evolve over time.

Despite the differences outlined above, the classical CSA model may nevertheless be conceptualized as being founded on four fundamental principles (Gorman, 2018; Henderson and Van En, 2007; Struś et al., 2020; Medici et al., 2023; Milford and Reed, 2024; Petruzzelli et al., 2023; DeMuth, 1993), which are synthesized in Table 1.1.

Connecting farmers and consumers (members) through long-term partnerships, including advance payments to cover production costs, is a fundamental principle of CSA (Cone and Myhre, 2000; Volz et al., 2016). In this way, both risks and harvests are shared. CSA differs from other cooperative approaches—for example, community gardens, where cultivation is primarily carried out by unpaid and non-professional volunteers, or food cooperatives, where risk is not shared between producers and consumers and the products are not necessarily sourced directly from primary producers (Haack et al., 2020). This means that the CSA model is not limited to the mere exchange of agri-food products for financial benefits but also places strong emphasis on social bonds and solidarity between producers and consumers (Hinrichs, 2000; Dankar, 2021; Flora and Bregendahl, 2012).

In this regard, CSA involves not only the sharing of costs, risks, responsibilities, and values—such as financial contributions, labor, and other resources—but also the exchange of skills and work between producers and members (Van En, 1995). In other words, it establishes a model of collective farming in which producers and consumers collaborate to

Table 1.1 Core Principles of the CSA Model

Principle	*Description*
Partnership	The price is determined based on transparently estimated food production costs and takes into account both the farmers' labor and the financial capabilities of consumers.
Solidarity	The risks associated with food cultivation are shared by both parties—consumers support farmers during difficult periods (e.g., crop failures) while benefiting from surplus yields in abundant seasons.
Proximity	Shortening supply chains and supporting the local economy. Consumers have the opportunity for direct contact with farmers and to gain insight into the realities of food production.
Diversity	Support for sustainable agricultural practices, such as crop rotation, which enhance soil health and biodiversity.

Source: Author's own elaboration.

support one another and to share the risks and benefits associated with food production (Balázs et al., 2016). As a result, producers are provided with a stable income, while consumers receive fresh, high-quality products (Milford et al., 2025; Witzling et al., 2020; Zepeda et al., 2013a). One of the key advantages of CSA is the distribution of risk between farmers and consumers (Bernard et al., 2020). Farmers receive a portion of the payment in advance (a subscription), which enables them to make necessary investments. At the same time, they are guaranteed a market for their products, which are delivered to consumers according to pre-agreed terms and delivery arrangements (Struś et al., 2020).

Although the arrangements proposed within CSA appear financially advantageous for farmers, the literature also highlights concerns that they do not adequately safeguard farmers' economic interests. For example, Paul (2018) shows that CSA farmers in the United States do not earn enough to sustain themselves, as agreements with consumers—intended to secure a decent income—fail to account fully for the farmer's full-time labor. Although prices are "set freely and fairly," they often do not provide farmers with an adequate income (Samoggia et al., 2019). Likewise, Galt (2013) notes that because income generation is not the primary objective of CSA formation, farmers frequently feel compelled to supply food to members at the lowest possible cost. Moreover, contrary to the model's assumptions, CSAs in the United States often compete for members, which further depresses farmers' income and undermines the very rationale of CSA (Galt et al., 2016).

In addition, Wellner and Theuvsen (2018) argue that low farm income may result from miscalculations made during budget planning, underscoring the importance of careful advance planning. Paul (2018) also observes that risk-sharing between farmers and consumers typically functions only in the short term, whereas in the long run it is the farmer who ultimately bears the risk. Furthermore, Medici et al. (2021) note that in the case of small Italian CSAs, member fees cover day-to-day operating expenses but are insufficient to support investment.

It is also important to emphasize that CSA is a model primarily designed for small-scale farms, which are often unable to compete with large-scale producers. The latter typically operate within a "quantity paradigm," where productivity, efficiency, financial surplus, and economies of scale play central roles—often driven by centralization, market competition, and the domination of nature (Czyżewski, 2015).

CSA may therefore be viewed as an economic model that transcends the simple logic of profit maximization, aiming instead to strengthen local interactions between food producers and consumers (Kraiß and van Elsen 2009), thereby laying the foundation for Alternative Food Networks (Opitz et al., 2019; Renting et al., 2003). In particular, CSA offers the potential to generate synergies between ecological, economic, and social dimensions, contributing to the transformation toward more sustainable food systems (Nost, 2014; Bloemmen et al., 2015; Birtalan et al., 2020).

The core values that define CSA can be summarized in nine key points (Volz et al., 2016; Cone and Myhre, 2000; Cristiano et al., 2020; Haack et al., 2020; Cristiano, 2021; Matzembacher and Meira, 2019; Struś et al., 2020; Fomina et al., 2022; Espelt, 2020):

1. Caring for soil, water, seeds, and other commons through the application of agri-environmental practices.
2. Treating food as a common good rather than a commodity.
3. Producing food on a scale that prioritizes human well-being and is rooted in local knowledge and conditions.
4. Ensuring fair working conditions and appropriate remuneration for all individuals involved in food production.
5. Respecting the natural environment and promoting animal welfare.
6. Ensuring universal access to fresh, local, seasonal, healthy, and diverse food.
7. Fostering community through direct and long-term relationships based on shared responsibility, risk, and benefits.
8. Encouraging engagement grounded in trust, mutual understanding, respect, transparency, and cooperation.
9. Promoting mutual support and solidarity that extends beyond national borders.

Based on the conducted analysis, it can be concluded that the CSA model provides a meaningful response to the challenges of contemporary agriculture by offering farmers financial stability and promoting ecological farming practices. At the same time, it enables consumers to access healthy, fresh, and local food directly from producers—strengthening social bonds and shortening food supply chains.

One may therefore argue that in the case of CSA, it is the social mechanism, rather than the price mechanism, that shapes market dynamics (Gruber, 2020). In mature economies, CSA members typically make collective decisions regarding the types of products, cultivation methods, and local distribution channels, and these decisions often take into account moral and ethical considerations (Wellner and Theuvsen, 2016, 2018). Solidarity among members plays a fundamental role, as it enables the support of core values such as the responsible use of resources, fair conditions for all parties involved, seasonal and local production based on agroecological practices, transparency, communication, and personal interaction (Carlson and Bitsch, 2019). Unfortunately, as will be demonstrated later in this monograph, in Poland—as well as in other post-socialist economies—the principle of solidarity is insufficiently appreciated and, in fact, constitutes a barrier to the development of CSA.

To summarize, despite the range of individual factors that influence farmers' and consumers' decisions to engage in CSA, the overarching motivation lies in the belief that industrial agriculture is unable, on the one hand, to provide consumers with high-quality food and, on the other, to ensure farmers a decent income (Carlson and Bitsch, 2019). Consequently, the shared aim of the CSA movement is to develop an alternative mode of shaping producer–consumer relations—one that facilitates direct exchange and enables a more sustainable system. In other words, CSA represents an alternative approach to food production, and perhaps even an alternative economic system, grounded in a critique of the current agri-food regime.

It is also important to note that CSA is gaining increasing recognition in policy-making, civil society, and academic discourse as a sustainable alternative to conventional food production systems. This model proved particularly relevant during the COVID-19 pandemic, when interest in local and more resilient food supply systems grew significantly (Stephens et al., 2020; Enthoven and van den Broeck, 2021).

Although the primary focus of this discussion is on CSA, it should be emphasized that CSAs constitute part of the broader category of Alternative Food Networks (AFNs). The origins of AFNs can be traced to consumer resistance against the standardization of agricultural production associated with the rise of industrial agriculture, as well as to the erosion of ties between producers and consumers. AFNs may even be seen as a response to growing public concerns regarding the social and environmental dimensions of food.

In the literature, AFNs are defined in various ways, reflecting their local embeddedness. Nevertheless, certain common elements can be identified, including the non-economic motivations of network participants, the belief in the high quality of products, the local nature of production, the importance of social relations, and an appeal to the principles of sustainable development (Goszczyński et al., 2018). The essence of AFNs appears to lie in their networked character and in the shortening of the chain of production, distribution, and consumption. In this broad understanding, AFNs encompass not only CSA but also all forms of direct food distribution, consumer cooperatives, producer groups, farmers' markets, urban gardening, allotment gardens, and food fairs.

A similar perspective is offered by Bilewicz and Śpiewak (2019), who argue that Alternative Food Networks may include a wide range of initiatives of different scales and characteristics, centered around the production, distribution, and consumption of food outside the conventional food system. In practice, most AFNs aim to reduce the uncertainties stemming from the absence of direct farmer–consumer relationships—a consequence of the development of industrial agriculture and the extension of food supply chains. To address this, AFNs seek to re-establish connections between consumers and local food producers, thereby fostering both economic linkages and social bonds. Alternative Food Networks also contribute to increased civic engagement and to strengthening the position of consumers.

At the same time, there is a noticeable tendency for AFNs to be incorporated into the mainstream agri-food market. This occurs, among other reasons, because retail chains have begun to include "alternative" products (organic, traditional, and farmer-sourced) in their assortments (Bilewicz and Śpiewak, 2019). Moreover, some grassroot AFNs undergo processes of commercialization (Brunori et al., 2012).

In summary, while neither CSA nor AFNs are likely to replace conventional industrial agriculture—which pursues increased productivity and economies of scale—they nevertheless offer viable opportunities for the survival of small farms that are unable to compete on price. CSA and AFNs may also serve as a means of diversifying production in larger agricultural holdings. Furthermore, they can strengthen local food security, the importance of which increased significantly following the pandemic, which exposed the vulnerabilities of long supply chains.

1.2 CSA versus Traditional Models of Food Production and Distribution

Although alternative models of food production and distribution are gaining increasing popularity, industrial agriculture—characterized by long food supply chains—continues to dominate on a global scale. This dominance can be attributed to the ongoing processes of globalization and the concentration of land and capital, both of which favor the expansion and influence of large corporations. As a result, agriculture has undergone a radical transformation—from the "peasant model" to the "mega-enterprise" farm. This form of intensive and concentrated agriculture, which currently shapes the European and American agricultural sectors, cannot be regarded as a future-oriented model, as it compels farmers to redefine their relationship with nature.

While industrial agriculture offers several advantages—most notably increased productivity—it is frequently criticized for a range of issues:

- **Economic**: the monopolization of food production by a small number of large, specialized farms, which may contribute to rising food prices;
- **Environmental**: soil degradation and water pollution, along with reduced biodiversity in agricultural ecosystems;
- **Socio-cultural**: the decline of small farms, leading to the erosion of traditional rural landscapes.

The key characteristics and impacts of industrial and sustainable agriculture are concisely summarized in Table 1.2.

Table 1.2 Characteristics and Impacts of Industrial and Sustainable Agriculture

Industrial Agriculture		*Sustainable Agriculture*	
Characteristic	*Impact*	*Characteristic*	*Impact*
Concentration	• Increased productivity and reduced production costs due to economies of scale. • Marginalization of small-family farms. • Monocultures increase the risk of plant diseases and soil degradation.	**Multifunctionality**	• Diversification of farmers' income sources (e.g., agritourism, direct sales). • Preservation and protection of agricultural traditions and rural landscapes. • Increased farmer engagement in environmental protection and local development.
Specialization	• Improved production efficiency and output. • Greater vulnerability to price fluctuations and market conditions. • Decline in biodiversity and reduced ecosystem resilience.	**Sustainability**	• Reduced soil degradation and environmental pollution. • Greater ecosystem resilience to climate change and plant diseases. • Preservation of long-term agricultural productivity.
Intensification	• Increased food production and availability. • Soil degradation and depletion of water resources. • Increased use of energy and fossil fuels.	**Family Farming**	• Strengthening of social bonds and the stability of local communities. • Greater care for product quality and land stewardship. • Limited ability to compete with large-scale industrial farms.
Chemical Inputs	• Short-term increase in agricultural productivity. • Contamination of soil and groundwater. • Negative impacts on human and animal health and pollution of ecosystems.	**Organic Farming**	• Production of high-quality, chemical-free food. • Lower soil and water contamination. • Lower yields compared to industrial farming, which may lead to higher product prices.

Source: Author's own elaboration.

The CSA model discussed in the previous section meets the criteria of sustainability, as it functions similarly to organic or family farming systems, which are generally characterized by greater environmental friendliness, economic viability, and social acceptability. These types of farms are typically embedded in the local community. In contrast to large-scale monocultural farms, CSA operations tend to be more environmentally conscious and more attuned to issues of sustainable development. They take into account the long-term consequences of their practices—for instance, ensuring that the farm is left in optimal condition for the next generation. By contrast, mega-farms are usually managed by individuals whose primary goal is short-term profit maximization, often at the expense of the environment.

Therefore, it can be argued that industrial agriculture provides benefits primarily to a shrinking group of large-scale farms, while increasingly disconnecting farm viability from the vitality of rural communities (Zegar, 2012, p. 36). On the other hand, it should be acknowledged that the productive capacities of smallholder farming in developing countries are often insufficient. Such farms frequently fail to secure adequate financial income and are unable to meet the growing demand for agri-food products among the population. Therefore, despite the critique of industrial agriculture presented above, it must be clearly emphasized once again that neither CSA nor, more broadly, AFNs will be able to replace it in the foreseeable future. Although authors such as Gliessman (2016), Méndez et al. (2013), and Mert-Cakal and Miele 2020) argue that the current system of food production and distribution should evolve in the direction envisioned by CSA, this process should be understood as a gradual evolution rather than a radical transformation. A radical shift—were it even feasible—would likely result in a food crisis and in a weakening, rather than a strengthening, of food security.

It should therefore be assumed that CSA and the broader AFN system will continue to operate in parallel with industrial agriculture, gradually gaining in significance. CSA will strengthen local food systems, particularly because it relies on short supply chains and is therefore more resilient to external disruptions such as wars or pandemics.

From a market structure perspective, agriculture was until recently associated with the model of perfect competition (Kamerschen et al., 1991), characterized by very low barriers to entry and a lack of price control. This market model was typical of fragmented agriculture, which supported sustainable development. Today, however, industrial agriculture increasingly resembles a quasi-monopoly. In such a structure, the entry of new participants is restricted, and price control is significant. Under monopoly conditions, a single enterprise can use its market power to set higher prices and reduce output relative to a competitive market, thus generating profits in both the short and long term.

Monopoly is also associated with increasing returns to scale, meaning that the average long-term cost per unit decreases as the production volume grows. As a result, buyers pay higher prices for smaller quantities of goods, since the monopolistic price exceeds the equilibrium price. Furthermore, monopolists may set different prices for different buyers, thereby eliminating competition from small-scale producers. This practice is known as price differentiation or price discrimination. In cases of imperfect price discrimination, monopolists often apply quantity-time discrimination, where prices vary depending on the volume purchased (e.g., lower prices for wholesale buyers). The monopolist's profits are protected by strong entry barriers, which in agriculture are primarily related to ownership and control over essential resources—especially land and water (Raftowicz, 2022, p. 30).

In contrast to a monopolist, a small-scale food-producing farm operating under conditions of perfect competition cannot offer prices higher than the prevailing market rate, as doing so would result in the loss of customers. Nor can it set different prices for different buyers. Small-scale agricultural producers also possess significantly less bargaining power compared to monopolists. This is largely due to the fact that, under monopoly conditions, transaction costs are considerably lower (Williamson, 1998). These include the costs associated with identifying business partners, designing, negotiating, and securing contracts (ex-ante costs), as well as monitoring contract implementation (ex-post costs). Transaction costs are strongly influenced by the institutional environment.

Small producers may attempt to reduce these costs through mutual cooperation; however, in some countries, such cooperation is hindered by a low level of social capital and concerns over free-riding behavior.

The development of industrial agriculture has also been driven by shifts in demand. Notably, the emergence of large retail chains and discount stores—functioning as monopsonies—has led to increased demand for large volumes of standardized products.

Alongside the expansion of industrial agriculture, food supply chains have also become significantly longer. In classical terms, the supply chain can be understood as a sequential flow from agricultural production, through food processing, wholesale and retail trade, and finally to the end consumer.

Long food supply chains are typically presented in context and contrasted with short supply chains. They are most commonly associated with a logistics system oriented toward production—particularly mass production—based on a strategy often referred to as a push strategy (Szymanowski, 2006, p. 36), in which existing market needs are met by pushing goods forward through the supply chain. A key distinguishing feature between these two types of supply chains is the number of intermediaries involved.

Long supply chains involve the mass distribution of products that pass through multiple intermediaries—such as wholesalers, wholesale markets, retailers, or processors—whereas short supply chains primarily focus on direct sales (either in person or online) to final consumers (Malak Rawlikowska et al., 2019).

The issue of long food supply chains is also closely linked to the concept of food miles, defined as the distance between the place where food is grown or produced and the place where it is consumed (Kemp et al., 2010). It is generally accepted that the longer the supply chain, the more kilometers the food must travel.

Due to the rapid development of transportation, low fuel prices, and growing demand for cheap, exotic agri-food products, long food supply chains continue to expand (Raftowicz, 2022, p. 54).

An important element in the definition of long food supply chains is their negative connotation, often associated with critiques of global, mass, or industrial distribution systems, and the belief that they have a harmful environmental impact compared to alternative, more ecologically sustainable methods of food distribution and consumption (Ilbery & Maye, 2005b; Gallaud and Laperche, 2016; Le Velly and Dufeu, 2015). This is further reinforced by the presence of negative environmental externalities generated by conventional food systems (Banks and Bristow, 1999), as well as the limited control farmers have over these chains (van der Ploeg & Frouws, 1999).

Analyzing the origins of long food supply chains, it becomes clear that they have existed in the public sphere for a relatively short period. Throughout most of human history—nearly 99% of it—agriculture and food needs were met through traditional means, with direct contact between producers and consumers (Chaturvedi et al., 2014).

Over the centuries, the development of market economies led to profound socio-economic transformations. Geographic discoveries also played a significant role, permanently reshaping international trade (Sołdaczuk and Misala, 2001, p. 51) and contributing to the lengthening of supply chains. Labor specialization, ongoing technological advancement, and the growth of transportation further weakened local markets.

The post–Second World War period marked a turning point. In highly developed countries, direct sales of agricultural and food products began to decline steadily. Starting in the 1960s, as developed nations rapidly industrialized, the structure of food supply chains began to shift. With the advent of the 21st century, the growing concentration of land and capital, along with mechanization, agricultural specialization, and the rise of large-format retail stores and supermarkets, led to the near disappearance of direct-to-consumer sales. These were replaced by long food supply chains and the expansion of super- and hypermarkets.

Economic integration of countries (especially in Europe) and globalization also played a key role in accelerating the development of extended supply chains. While globalization has undoubtedly brought significant benefits to the food trade—such as increased exports, job creation, and improved access to technology, products, and services—it also has negative implications. Among them are the growing commodification of agriculture and food, and the rise of global commodity chains (Gereffi et al., *2005*). In this context, retail chains and global food brands often take control over the flow of goods throughout the entire chain, imposing both prices and quality standards on other actors.

Paradoxically, the lengthening of food supply chains has generated growing concerns about food security. For many years, this problem went largely unnoticed; on the contrary, it was often assumed that globalization and the international flow of food helped to improve food security. The Food and Agriculture Organization of the United Nations (FAO, 2021) defines food security as a situation in which all people, at all times, have physical and economic access to sufficient, safe, and nutritious food that meets their dietary needs and food preferences for an active and healthy life.

Unfortunately, the current state of food security has markedly deteriorated as a result of the pandemic, climate change, economic disparities, and ongoing armed conflicts (Mansura, 2024). As noted by Onyszkiewicz and Sylla (2025), on average, one in four people worldwide experiences moderate or severe food insecurity. It has become evident that long supply chains are highly vulnerable to disruptions—an issue that became particularly visible during the COVID-19 pandemic. During this period, restrictions on production, processing, and trade led to a decline in food production. It should be stressed that the problem concerned not only the production and transportation of food but also seeds and fertilizers. As a consequence, food prices increased.

In conclusion, the intensifying and often unregulated processes of globalization, the concentration of land and capital, fierce price competition, and the industrialization of agriculture have contributed to the emergence of an unsustainable agricultural model. This imbalance is now manifested in the dominance of industrial farming and food distribution based on long supply chains. Moreover, there is a feedback loop between the concentration of land and production and the continued lengthening of supply chains. Although the complete elimination of industrial agriculture is not feasible, its transformation is both necessary and possible—through the involvement of appropriate institutions that can foster sustainable development in the agricultural sector. This includes the promotion of more localized, ecologically sustainable, and socially responsible models. Only in this way can a long-term balance be achieved between production efficiency, the preservation of natural resources, and social well-being.

1.3 The Functioning of CSA in Selected "Old" EU Member States

As mentioned earlier, the idea of CSA reached Europe in the late 1970s, first appearing in Switzerland and subsequently spreading to France, Germany, and Italy. Today, the majority of European CSA initiatives are concentrated in France, with over 2,000 initiatives (Volz et al., 2016; Egartner et al., 2020). In Western Europe, the model has taken on a character distinct from that in the United States. Unlike the American version of CSA, which tends to be more transactional in nature (Paul, 2018), European initiatives place greater emphasis on collective co-production and the principle of solidarity. In Europe, consumers often actively participate in cultivation and farm management, and the entire system operates on principles of social justice, shared risk, and equitable access to high-quality food (Devere and Lamine, 2010).

In the European literature, these ideas were further developed by Renting et al. (2012), who describe CSA and other Alternative Food Networks as manifestations of so-called food democracy. According to the authors, civic movements around food represent a form of "reclaiming" control over the food system by citizens through the co-creation of new forms of participation, trust, and social responsibility. In this perspective, community-based models are interpreted as civic food networks, which link issues of production and consumption with social, environmental, and ethical values. The authors emphasize that building food democracy requires not only transforming market relations, but also strengthening social capital and community-based participatory practices.

These ideas found practical expression in the development of community movements in various European countries, particularly in France, where the AMAP (*Association pour le Maintien d'une Agriculture Paysanne*) emerged and quickly gained popularity, inspiring similar initiatives across Europe. The first AMAP was founded in 2001 in Aubagne (southern France) by a group of farmers and members of the organization Attac. Since then, the number of AMAPs has grown exponentially. By 2015, more than 2,000 AMAPs were operating across all regions of the country, serving over 250,000 consumers who regularly collected food baskets from local producers under contractual agreements. In 2017, AMAPs generated an annual turnover of approximately €150 million (Philipon et al., 2017, p. 65). To this day, AMAP remains the most developed and organized form of CSA in Europe (Wezel et al., 2018). In 2020, the number of registered AMAPs reached 2,120 (URGENCI, 2021). This success is partly attributed to France's strong agricultural traditions and the deep cultural ties between agriculture, gastronomy, and tourism.

The AMAP model emerged as a response to the increasing industrialization of agriculture and the need to rebuild relationships between producers and consumers. Its primary objective is to support ecological and local farming, and to ensure a fair distribution of agricultural production costs and risks (Parot et al., 2024; Bashford et al., 2015). Since 2003, AMAPs have operated under a Charter of Values developed by Alliance Provence. This document outlines 18 key principles, including support for organic agriculture, fair pricing, and risk-sharing between farmers and consumers (Marquat and Diemer, 2016).

Both consumers and farmers play essential roles in the AMAP system. However, consumer-initiated models are the most prevalent—groups of consumers organize themselves and seek out farmers willing to cooperate. The goal of these initiatives is not only to ensure access to healthy, local food but also to support farmers facing market challenges (Moruzzi and Sirieix, 2014). AMAPs are based on long-term partnerships between farmers and consumers. Members commit to purchasing a predetermined quantity of products throughout the season, which guarantees farmers a stable income and eliminates the need for intermediaries. In return, consumers receive regular deliveries of fresh farm products—typically in the form of weekly food baskets (Parot et al., 2024).

An AMAP consists of a group of consumers and one or more producers. This collaboration is governed by a long-term agreement and is built on several key principles (Bashford et al., 2015; Schlicht et al., 2012; Marquat and Diemer, 2016):

- **Direct contracts**: farmers and consumers enter into agreements for a defined period (e.g., the growing season).
- **Advance payments**: consumers pay in advance for a specified number of food deliveries, providing financial stability for farms.
- **Risk-sharing**: in the event of a poor harvest, consumers accept smaller deliveries instead of placing the entire burden on farmers.
- **Social engagement**: AMAP participants often assist with farm activities through hands-on work or by helping organize distribution.

The organizational structure of AMAP is decentralized—each group operates independently and may differ in specific operational rules. However, umbrella organizations such as MIRAMAP (*Mouvement Inter-Régional des AMAP*) coordinate network development at the national level (Parot et al., 2024).

In France, legal frameworks have been introduced to promote the use of local food in schools, hospitals, and other public institutions. These policies help AMAP farmers reach broader consumer groups while also advancing sustainable agriculture and strengthening local economies. Such regulations offer vital support for initiatives like AMAP, where public procurement becomes a tool for ensuring stability and stimulating demand for farm products tied to this model (Légifrance, 2018).

Despite its many advantages, the AMAP system faces several challenges that may limit its future expansion (Schlicht et al., 2012; Marquat and Diemer, 2016; Parot et al., 2024):

- Relatively low public awareness: many people are still unfamiliar with the model, and AMAP groups often lack resources for large-scale promotion.
- Membership costs: for lower-income individuals, the upfront payment may be a barrier, although some AMAPs offer flexible payment options or discounts.
- Limited product variety: members receive seasonal produce, which may not meet the expectations of those used to the broad selection available in supermarkets.
- Collection obligations: the need to collect food regularly at fixed times and locations may be inconvenient.
- High organizational demands: managing an AMAP requires active member involvement, which can discourage those looking for simpler food purchasing options.
- Structural diversity and lack of standardization: not all AMAPs are officially registered within the MIRAMAP network, complicating coordination with local authorities.
- Varying levels of consumer engagement: for some participants, AMAP is merely a source of organic food rather than a form of social activism.
- Economic limitations for farmers: although the model provides financial stability, not all farms can afford to rely exclusively on this sales method.
- Urban development pressure: many AMAPs operate in urban or peri-urban areas where expansion threatens farmland. Urban sprawl consistently weakens agricultural viability and creates conflicts between development and farmland preservation. For example, in France's Bouches-du-Rhône region, approximately 1,000 ha of farmland are lost each year to urbanization (Gallaud and Laperche, 2016).

In conclusion, the French AMAP model is one of the most dynamic initiatives supporting local and ecological agriculture in Europe. Through the active involvement of both consumers and farmers, the system enables fair distribution of production costs and risks while fostering long-term relationships between producers and consumers. Despite its challenges, AMAP remains one of the most successful models of cooperation within the alternative agriculture sector, offering access to healthy, local food and supporting sustainable rural development. It is also worth emphasizing that public procurement regulations in France play a key role in sustaining AMAPs by prioritizing local food systems.

The development of CSA in Germany followed a somewhat different trajectory. While the model has been present for several decades, its significant rise in popularity has occurred only in recent years. The first CSA initiative in Germany was launched in 1988 in Lauenburg. It emerged as a response to growing dissatisfaction with conventional agriculture and the desire to create an alternative model for accessing food. Despite these early experiments, CSA remained a niche concept for a long time. In the following 15 years, only three CSAs were established, and between 2003 and 2007, their number increased to eight (URGENCI, 2021). This was due to several factors (Meyer and Hassler, 2023; Bonfert, 2022):

- a lack of sufficient knowledge about alternative forms of agriculture;
- the dominance of supermarkets and conventional food supply systems;
- a limited number of farms willing to transition to the CSA model.

Since 2010, CSA—known in Germany as *Solidarische Landwirtschaft* (Solawi)—has been expanding, focusing on supporting sustainable agriculture and the protection of small farms. In 2015, there were 78 CSA farms; by 2022, the number had grown to 398, with an additional 91 in the process of formation (Meyer and Hassler, 2023). By 2024, there were approximately 500 initiatives operating across the country (Middendorf and Rommel, 2024). The reasons behind this dynamic growth include:

- growing ecological awareness and consumer social responsibility;
- the climate crisis and the search for more sustainable food production methods;
- the COVID-19 pandemic, which increased interest in local and direct food sources;
- organizational support from networks such as *Netzwerk Solidarische Landwirtschaft*, which assists with the establishment of new initiatives.

Currently, CSA is present across all regions of Germany, although it initially gained a foothold in the northern part of the country.

A key feature of German CSA is its organizational structure. Two main models of organization can be identified (Meyer and Hassler, 2023; Bonfert, 2022):

- Consumer-driven farms: community members are actively involved in farm management and decision-making, forming working groups, organizing meetings, and hosting educational events. Participants may contribute to decisions about production and engage in agricultural work. Consumers purchase "shares" in the farm and regularly receive a

portion of the harvest. This model often operates on a voluntary contribution basis, enabling access to food for people of varying income levels. In addition to purchasing food, Solawi members often engage in farm work, environmental education, and organizing local events.
- Producer-driven farms: farms are managed by farmers themselves, while consumers primarily act as financial supporters.

Considering the way CSA initiatives are managed, they can be divided into three models (Nascent (2023):

Type 1: Producer-led CSA. Bilateral individual contracts between the producer and consumers.
Type 2: Cooperative CSA. Cooperation between independent producer(s) and a legally organized member cooperative.
Type 3: Co-entrepreneurship CSA. Joint cooperation between producers and consumers within a single legal entity.

Studies conducted on 22 CSA farms in the Hesse region show that the majority follow the consumer-driven model, where community participation is essential for the farm's functioning (
Meyer and Hassler, 2023).

The benefits of this model include:

- Increased ecological awareness: education about seasonality, certification, and the real costs of food production;
- Local community development: joint events, member gatherings, and cooperation between farms;
- Sustainable food systems: reducing dependency on global supply chains and supporting small-scale farmers.

Cooperation between farms takes place on various levels:

- Local: collaboration networks between neighboring CSA farms, such as those around Leipzig, where several initiatives work together;
- Regional: regional groups such as *Mittendrin* in central Germany, which support the development of new initiatives;
- National: the *Netzwerk Solidarische Landwirtschaft* provides advisory services, educational materials, and legal support for new CSA projects.

As the number of CSA farms has grown, new challenges have emerged as well (Bonfert, 2022; Wittenberg et al., 2022):

- Increased competition among farms: more CSA projects mean competition for members and financial resources;

- Lack of skilled labor: difficulties in finding experienced gardeners and farmers;
- Uneven levels of member engagement: not all consumers actively participate in community life, which may weaken the cooperative spirit.

Research has shown that CSA is not suitable for every farm or consumer in Germany, meaning that universal solutions cannot be applied given the regional and structural diversity of the farming sector. Moreover, there is a need to avoid the homogenization of agricultural methods, which is considered one of the drivers of the agricultural crisis. Despite its advantages, CSA must also be adapted to other production systems, including larger-scale farms. Another challenge is the need for ongoing research and broader dissemination of the CSA concept among consumers and policymakers, in order to fully unlock its potential.

Despite these challenges, CSA in Germany continues to grow, offering a sustainable alternative to conventional food systems and fostering strong, locally rooted communities around agriculture.

The development of CSA in Italy followed a somewhat different path. In 1986, Carlo Petrini founded the Slow Food movement in Italy as a response to the growing homogenization of the food system and the accelerating pace of life, both of which favored mass food production and the expansion of fast-food culture. *Slow Food* promotes traditional methods of food cultivation, production, and preparation, encouraging conscious consumption and the protection of local products (Ilbery and Maye, 2005a; Pirog, 2009). The movement was officially institutionalized in 1989 with the signing of the Slow Food Manifesto in Paris, establishing its core values: taste, quality, locality, and sustainability. Since then, the organization has expanded globally, supporting biodiversity, traditional production methods, and local farming.

Slow Food also played a crucial role in the emergence of alternative food systems in Italy, including Gruppi di Acquisto Solidale (GAS) and CSA. GAS consists of consumer groups who collectively purchase goods directly from farmers, eliminating intermediaries. Their main characteristics include:

- Organization: they operate as cooperatives, relying on local community collaboration;
- Philosophy: they emphasize social justice, environmental protection, and support for small, family-run farms;
- Product diversity: beyond food, GAS groups often include ecological clothing or household products;
- Impact: thanks to strong local social networks, this model fosters the growth of a sustainable economy.

Alongside GAS, CSA is another important form of community-based agriculture in Italy.

According to research by Piccoli et al. (2021), Italian CSA initiatives are typically producer-driven: of the 16 cases analyzed, 12 were initiated by farmers, while only 4 originated from consumer groups. The organizational structure may involve single producers, clusters of up to six cooperating farms, or larger multi-farm organizations. Most Italian CSA initiatives do not have a formal legal status—only 5 out of the 16 analyzed had clearly defined operating rules. This model is often based on trust-based relationships between producers and consumers, which form the foundation of cooperation.

Italian CSA initiatives usually focus on vegetable production, and the farms involved tend to be relatively small in size—half operate on plots of up to 1 ha, and only a few manage more than 8 ha.

Despite their many advantages, CSA initiatives in Italy face several challenges that hinder their further development. The main difficulties include (Medici et al., 2021; Rossi et al., 2024; Piccoli et al., 2021; Fonte, 2013):

- Low consumer engagement: many participants contribute only financially without taking part in farm work, weakening the sense of community;
- Lack of consumer loyalty: some members participate for just a year, making it difficult to ensure long-term stability;
- Farmers' financial difficulties: challenges in setting fair remuneration lead to self-exploitation and reduced motivation among producers;
- Legal barriers: labor regulations, including those concerning volunteer work, reduce operational flexibility;
- Lack of public recognition: CSA initiatives often do not receive institutional support, limiting their ability to grow;
- Economic challenges: most CSA initiatives are financed primarily through member contributions, making them vulnerable to market fluctuations and limiting their ability to meet the long-term environmental and social goals typical of CSA.

Nevertheless, Italian CSA remains a dynamically evolving model of alternative agriculture, redefining relationships between producers and consumers. Although still relatively new, these initiatives bring forward innovative and sustainable approaches to food production, grounded in principles of social and ecological justice. In the face of structural and institutional challenges, they continue to demonstrate the potential of cooperative, community-based agriculture (Mastronardi et al., 2019). Currently, approximately 2,000 CSAs are registered in Italy (URGENCI, 2021).

In summary, CSA initiatives in the so-called old EU Member States not only take diverse forms but share a common goal: supporting sustainable agriculture, strengthening social ties, and ensuring access to high-quality food for consumers. Examples such as France's AMAP, Germany's Solawi, and Italy's GAS and CSA initiatives illustrate the effectiveness of this system in transforming agriculture and rural areas, while enhancing ecological awareness throughout Western Europe.

1.4 CSA in Central and Eastern Europe—Current Practices and Perspectives

Community-Supported Agriculture in Central and Eastern Europe—particularly in the Czech Republic, Hungary, and Romania—has evolved along a distinct trajectory compared to Western European countries. Although the conceptual foundations of CSA are broadly similar, encompassing direct farmer–consumer partnerships, the sharing of risks and benefits, and the promotion of local ecological production, its practical implementation in this region deviates from Western European models (Moellers and Bîrhală, 2014). This divergence is reflected in the comparatively low number of CSA initiatives, as presented in Table 1.3. The limited prevalence of such initiatives can be attributed to the structural characteristics of local agricultural systems and the historically shaped nature of producer–consumer relationships.

In the Czech Republic, CSA has been developing dynamically since around 2010. A national network, *Asociace místních potravinových iniciativ* (AMPI), was established to support the organization of CSA initiatives and foster the development of short food supply chains, as well as to serve as a platform for the exchange of experiences between farmers and consumers (ARC, 2020).

Czech CSA farms are characterized by solid organizational structures, a growing number of participants, and a relatively high level of consumer

Table 1.3 Number of CSA Initiatives in Central and Eastern European Countries

Country	*2016*	*2020*
Czech Republic	23	80
Slovakia	10	Data not available
Hungary	8	Data not available
Poland	8	8
Romania	5	10

Source: Author's own elaboration based on Overview (2016) and URGENCI (2021).

awareness. As a result, these farms tend to operate more stably and show greater potential for long-term development (Overview, 2016).

As noted by Jehlička and Smith (2011), the development of this model is closely linked to the long-standing tradition of self-provisioning and household-level food production in Czech society, practices that date back to the socialist period. These activities did not disappear after the political transformation; instead, they evolved into modern forms of collaboration within local food networks. This phenomenon aligns with the concept of "quiet sustainability"—everyday, informal practices grounded in self-sufficiency, reciprocity, and local social ties (Smith and Jehlička, 2013).

Research by Vávra et al. (2017) shows that nearly half of Czech households produce part of their own food, which strengthens the resilience of local food systems and supports the acceptance of alternative models such as CSA. As Jehlička et al. (2019) observe, practices of household gardening and neighborly exchange constitute a form of "everyday resistance" to dominant models of production and consumption, reinforcing local communities and household autonomy. In this context, CSA in the Czech Republic has been developing not only as an alternative economic model but also as an element of social and cultural transformation, integrating rural and urban residents around the values of self-sufficiency and trust.

Goszczyński (2019) emphasizes that the Czech experience demonstrates how alternative food systems in Central and Eastern Europe often develop outside formal market structures, and how participants do not always perceive their activities as ecological or socially oriented, even though they contribute to the sustainable development of rural areas.

Moreover, URGENCI (2021) reports that Czech CSA initiatives are among the most stable in the region: during the COVID-19 pandemic, they maintained supply continuity, increased membership, and expanded their educational and community-oriented activities. Cooperation within the AMPI network during the health crisis further strengthened trust between farmers and consumers, underscoring the growing importance of social capital as a key factor in the resilience and long-term development of short food supply chains in the Czech Republic.

In Hungary, the first CSA initiatives emerged after 2010, inspired by Western models. Although the number of operational CSA farms remains relatively small, some—such as *Nyílt Kert* and *Zsámboki Biokert*—have established a stable presence and gained local support. Hungarian CSA focuses primarily on organic production and short supply chains. However, its development is hindered by a lack of institutional support and relatively low public awareness of alternative food systems (Balázs et al., 2016; Samoggia et al., 2019).

In the early 2010s, the first grassroots local food systems began to emerge in cities such as Gödöllő and Szekszárd, as demonstrated in the research by Balázs (2012). These initiatives connected citizen groups, researchers, and local governments and served as precursors to today's CSA schemes, representing a "second generation" of Hungarian short food supply chains that developed in response to institutional changes after 2010.

At the same time, within the framework of the Rural Development Program 2014–2020, the Short Supply Chain Thematic Programming subprogram (REL) was launched to support small producers in local sales and in the promotion of high-quality food (Bakos and Khademi-Vidra, 2019). The buying groups and producer networks established under this program not only strengthen local economies but also foster community integration and contribute to the growth of culinary tourism.

As in the Czech Republic, traditions of self-provisioning and local exchange have played an important role in the development of alternative food systems in Hungary. According to Vávra et al. (2017), a significant share of Hungarian households continues to produce food for their own consumption, which helps build local reciprocity networks and facilitates the acceptance of models such as CSA. Goszczyński (2019) further observes that in post-communist countries, including Hungary, many food production and distribution practices remain informal and grassroots in nature, yet they nonetheless contribute to sustainable development and to strengthening social resilience.

According to URGENCI (2021), Hungarian CSA groups are among the most organizationally diverse in the region. Alongside classical farmer–consumer partnerships, hybrid initiatives have emerged that combine direct sales, short food supply chains, and urban community gardens. These models promote rural–urban integration and enhance food system resilience in times of crisis, such as during the COVID-19 pandemic.

Balázs et al. (2016) emphasize that the development of local food systems in Hungary is driven primarily by civic initiatives and urban consumer groups, which reduce the distance between producers and consumers and fill gaps left by limited public support. Benedek et al. (2018) similarly note that despite the dominance of industrial agriculture following EU accession, consumer interest in local products and sustainable farming has been steadily increasing. As a result, emerging CSA groups and food cooperatives often take the form of bottom-up, flexible initiatives, although they may also be vulnerable to financial instability.

As highlighted by Kovách and Megyesi (2024), land concentration and the deagrarianization of society have weakened local social ties, meaning that Hungarian CSA systems operate under conditions of post-peasant society. Nevertheless, in recent years they have increasingly aligned with post-growth and food sovereignty perspectives. Initiatives such as the Budapest Food City Lab function as laboratories of social transformation, bringing together researchers, farmers, and urban residents to co-create solutions for sustainable food systems.

Despite numerous barriers, Hungarian CSA initiatives are becoming an important component in rebuilding trust between farmers and consumers and represent a significant form of social innovation in the rural areas of Central and Eastern Europe.

In Romania, CSA is far less developed than in the Czech Republic or Hungary. Although the first initiatives began to appear after 2010, overall growth has been slow, primarily due to low consumer awareness, limited institutional support, and farmers' continued reliance on traditional distribution channels (Overview, 2016). Most CSA farms operate near major urban centers, such as Bucharest or Cluj-Napoca, where demand for organic and local food is comparatively higher.

As Havadi-Nagy (2021) notes, alongside CSA initiatives, other forms of alternative food networks have been developing in Romania, including ASAT (Asociația pentru Susţinerea Agriculturii Țărănești), Cutia Țăranului, Bejön a vidék, Székely Termék, and a network of pilot food hubs supported by the Romanian-American Foundation. Although these initiatives operate on a small scale, they mobilize small producers and strengthen rural–urban linkages. Many farmers continue to rely on traditional distribution channels, which limits the potential for scaling these models (Overview, 2016). Furthermore, weak producer cooperation, low investment capacity, and the limited geographical reach of initiatives mean that AFNs in Romania remain largely niche phenomena (Möllers et al., 2018).

The first fully functioning CSA partnerships were established in 2009 in the Timişoara region through the efforts of the non-governmental organization CRIES (Centrul de Resurse pentru Inițiative Etice și Solidare), in cooperation with the French URGENCI network. The ASAT model was grounded in principles of solidarity, trust, and shared responsibility between farmers and consumers. As emphasized by Bîrhală and Möllers (2012) and Möllers and Bîrhală (2014), CRIES played a key role in forming producer–consumer groups and in running campaigns promoting local food ("Eat Healthy and Support Local Agriculture"). Over time, the organization transferred responsibility to participants, helping to establish self-sustaining and durable collaborative structures. The first ASAT farms were created by new entrants, often from outside the traditional agricultural sector, highlighting the importance of human capital and entrepreneurship in CSA development. The model relied on trust, solidarity, and shared responsibility, addressing inefficiencies within the agri-food market and offering an example of solidarity-based economics. By 2015, the ASAT concept had expanded to cities such as Arad, Oradea, Sibiu, Iași, and Bucharest, while remaining a niche initiative. Its strength lies in building trust and social capital around the idea of food justice, distinguishing Romanian CSA initiatives from the more commercialized forms found in Western Europe.

According to Tanasă et al. (2015) and Möllers and Bîrhală (2014), Romania's agrarian structure is highly fragmented, with approximately 3 million farms under 5 ha. While some scholars view this fragmentation as a barrier to development, others see potential for short food supply chains due to strong local ties and production flexibility. The greatest challenges include low mechanization, a lack of cold-chain infrastructure, difficulties in meeting sanitary standards, and weak producer cooperation. Additionally, around 48% of agricultural production is consumed within households, which limits the development of formal markets.

Despite these barriers, CSA/ASAT initiatives represent a form of social innovation that integrates economic, social, and environmental dimensions with the principles of solidarity-based economics. From the perspective of rural development policy, ASAT stands as an example of good practice in promoting sustainable development, and the Romanian CSA model—although still niche—serves as an important reference point for countries in Central and Eastern Europe.

As limited research indicates, key differences between CSA in Central and Eastern Europe and Western Europe relate to several crucial aspects. First, the scale and organization of CSA initiatives are significantly larger in the West. In France, Germany, or Italy, there are hundreds of well-established CSA farms with strong consumer bases. In contrast, in Central and Eastern Europe, CSA is still in a developmental phase, and smaller participant numbers make long-term financial stability more difficult to achieve.

Another significant factor is institutional support. In Western Europe, numerous organizations, foundations, and government programs actively support CSA development, making it easier for farmers to operate. In Central and Eastern Europe, such systems are largely absent, meaning CSA initiatives are typically grassroots and financially fragile (Volz et al., 2016).

The financial model also differs across regions. In Western Europe, CSA is often based on annual subscriptions or long-term agreements, providing farmers with greater financial predictability. In Central and Eastern Europe, payments are usually shorter-term—weekly or monthly—forcing CSA farms to seek new customers more frequently and reducing income security. Public awareness of environmental and social issues is also higher in the West, translating into greater interest in CSA and a stronger willingness among consumers to commit long-term to supporting farmers. In Central Europe, this awareness is still emerging, and many consumers are not yet ready to adopt this model of food purchasing.

Access to land and infrastructure is another notable difference. Western European countries tend to have better regulations ensuring access to land for smallholders, as well as infrastructure that facilitates CSA operations. In contrast, farmers in Central and Eastern Europe often face land ownership issues and a lack of appropriate tools and technologies for product distribution (Möllers and Bîrhală, 2014).

As noted by Goszczyński and Knieć (2011) and Goszczyński (2019), the development of CSA in Central and Eastern Europe is shaped by a social and historical context that differs markedly from that of Western countries, where such initiatives stem from mature consumer and environmental movements. In post-communist states, these initiatives tend to be more pragmatic in nature, often rooted in traditions of self-provisioning and informal exchange. Their development relies more heavily on social relationships, trust-based capital, and grassroots action than on institutional support.

In summary, CSA in Central and Eastern Europe is growing, albeit on a smaller scale than in the West. The absence of institutional support, limited social awareness, and unstable financial models constrain its expansion. Nonetheless, increasing interest in organic food and local distribution systems suggests strong potential for CSA to become a more widespread form of agricultural production and a viable alternative to conventional supply chains in the future (Struś et al., 2020).

Although the topic of CSA in Central and Eastern Europe is gaining importance, it remains underexplored in academic literature. Most studies focus on Western European and North American experiences, while the unique characteristics of CSA in post-communist countries—shaped by distinct historical and socio-economic contexts—require deeper analysis. There is a lack of comprehensive comparative studies that consider both farmer and consumer perspectives, as well as the influence of agricultural policy on CSA development.

Future research should focus on financial mechanisms, organizational effectiveness, the resilience of CSA models, and their impact on local communities and food systems. Such inquiry is essential to better understand the potential and challenges facing CSA in Central and Eastern Europe.

2 The Role of CSA in Selected Theories of Rural Development

2.1 The Neoliberal Development Paradigm

Socio-economic development is a complex process, and therefore no single universal theory can fully explain its mechanisms. In the second half of the 20th century, however, the neoliberal development paradigm gained a dominant position, significantly influencing the functioning of Central and Eastern European economies after 1989, including the trajectory of transformation in Polish agriculture and rural areas.

The neoliberal paradigm is based on two main theses of supply-side economics. First, production is considered the primary goal of the economic system, as it enables an ever-higher level of need satisfaction. Second, the market is seen as the most effective mechanism for the optimal allocation of resources (Ellison and Pino, 2012; Neilson, 2019; Venugopal, 2015).

At the heart of the neoliberal paradigm are theories of economic growth, often equated with increases in per capita GDP. From this perspective, growth is defined as the expanding capacity of an economy to produce goods and services—capacities which, in theory, have no upper limit. H. Binswanger (2012) even describes this as a "growth spiral" that continues to accelerate. The neoliberal paradigm assumes that the development process is endogenous, meaning it is driven by factors within the socio-economic system, and that any external threats to development can be overcome within this framework.

A central tenet of this paradigm is that all social barriers to development can be overcome through economic growth—specifically, through the increased productivity of production factors. According to W.W. Lewis, "higher productivity societies displace lower productivity societies. The former have been more successful in overcoming cultural, religious, ethnic, climatic, and political obstacles to productivity" (2005, p. 16). This assertion exemplifies a form of economic determinism.

DOI: 10.4324/9781003745198-3

The theoretical foundation of the neoliberal paradigm is rooted in the concept of bounded rationality, the theory of state failure, and institutional theory. The concept of bounded rationality suggests that social systems should generate material incentives, as these are the primary drivers of economic behavior. Market competition is regarded as the only effective mechanism for ensuring system efficiency. Meanwhile, the theory of state failure argues that government intervention disrupts the efficiency of market mechanisms. Many neoclassical economists believe that minimal state involvement is a prerequisite for well-functioning markets.

It is worth emphasizing that the market's effectiveness is shaped by a set of institutional constraints, including both formal rules (such as laws and regulations) and informal norms, as well as how these rules are enforced (North, 2014, p. 80). The influence of incentives on economic behavior thus depends on the institutional framework, which, according to D.C. North, must be organized in such a way as to stimulate productivity growth (North, 2014, p. 89). As such, market development depends on institutions that determine transaction costs. North also argues—perhaps less obviously—that the adaptive efficiency of economic systems is also institutionally determined. In his view, this efficiency is not dependent on the flexibility of supply and demand, but rather on "a set of institutions that are easily able to adapt to shocks, disruptions, and the ever-present uncertainty that characterizes every society at all times" (North, 2014, p. 81).

The selective relationship between neoclassical economics and institutional economics is evident in the preference for formal institutions, particularly individual property rights to scarce resources. In other words, the paradigm promotes privatization as the primary method of institutional change. At the current stage of development, the objects of privatization may include ownerless, common, or public goods. Notably, during the economic transformation of Central and Eastern Europe, this approach shaped how state-owned agricultural enterprises were privatized, which in turn led to increased land ownership concentration.

Within the neoliberal development paradigm, it is difficult to find room for the concept of Community Supported Agriculture. In fact, the Community-Supported Agriculture (CSA) model is largely at odds with the paradigm's core assumptions, particularly those concerning the productivity of production factors. Small-scale farms that operate extensively and rely heavily on manual labor are, from a neoliberal perspective, considered barriers to rural development and obstacles to the transformation of agriculture into an industrialized, market-oriented sector. The neoliberal model favors land consolidation and increased farm size, enabling the use of advanced machinery and the realization of economies of scale.

This results in increased capital productivity, achieved through more efficient resource use, and increased land productivity, enabled by larger-scale operations. Meanwhile, labor productivity is raised through the reduction of labor inputs, which in practice leads to a decline in rural employment and population outflow from the countryside.

In conclusion, under the neoliberal paradigm, CSA is viewed as an impediment to the transformation of Polish agriculture into an industrial model. It is seen as perpetuating a fragmented agrarian structure in rural areas and potentially reducing the international competitiveness of Polish agriculture. Accordingly, CSA is not considered a model that should be supported in Poland.

2.2 Critical Approaches to the Neoliberal Development Paradigm

In contemporary literature, the limitations of the neoliberal development paradigm are increasingly subject to critique, with scholars arguing that its focus on GDP growth and productivity fails to capture the real social and ecological costs of economic activity. Critical approaches emphasize that the neoliberal paradigm—equating development with capital accumulation and rising efficiency—contributes to growing inequalities, the weakening of social bonds, and the degradation of the natural environment. In this context, alternative development frameworks have gained prominence, including post-growth economics (Jackson, 2009; Raworth, 2017), the economics of the commons (Ostrom, 1990), and the theory of diverse economies (Gibson-Graham, 2006), all of which advocate for a plurality of economic practices, the sharing of resources, and the abandonment of profit maximization as the dominant logic.

From an economic perspective, a key critique of neoliberalism concerns its identification of development with economic growth, understood as an increase in GDP per capita. As Stiglitz et al. (2009) point out, this indicator does not reflect actual improvements in quality of life and often conceals rising social and territorial inequalities. In countries that adopted neoliberal reforms, a "growth paradox" frequently emerges: despite economic growth, there is no corresponding improvement in social well-being (Jackson, 2009). Piketty (2014) demonstrates that the logic of unlimited capital accumulation leads to increasing wealth concentration and the erosion of the middle class, which in the long run undermines social cohesion and the stability of democratic institutions.

Harvey (2005) argues that neoliberalism is not merely an economic doctrine but also a political project, one that restructures power relations through privatization, deregulation, and the marginalization of the state. As a result, neoliberalization generates systemic instability, concentrates profits in the hands of narrow interest groups, and deepens socio-spatial inequalities—particularly in peripheral and rural regions, which are increasingly reduced to resource bases for global markets.

From a social perspective, the neoliberal development paradigm has been criticized for promoting individualism, competition, and an instrumental approach to human relationships. As Sennett (1998) notes, the neoliberal culture of flexibility and professional success leads to an "erosion of character" and the disappearance of durable social ties. Similarly, Putnam (2000) observes that the emphasis on individual economic success, materialism, and the privatization of social life contributes to declining trust and the weakening of social capital—a key resource for the development of local communities.

In this context, Sandel (2012) formulates the concept of the "market society," in which the logic of economic exchange penetrates spheres previously insulated from market mechanisms—such as education, health, or culture. This process marginalizes values such as solidarity, care, and shared responsibility. Critics of neoliberalism therefore call for the restoration of the communal dimension of economic life.

In environmental terms, neoliberalism also exacerbates the climate crisis. Globalization, as its practical manifestation, promotes the privatization of natural resources, the expansion of transnational corporations, and the outsourcing of environmental costs to peripheral countries. This results in the "marketization of nature," whereby the environment becomes an object of speculation and a source of profit rather than a common good. As Raftowicz (2020) points out, uncontrolled globalization reinforces the cult of unlimited economic growth and has become one of the major drivers of accelerating climate change. This highlights the need for a transition toward a post-growth model that prioritizes international cooperation and shifts from profit maximization to the protection of natural resources and the quality of life of future generations.

The most radical critique of neoliberalism comes from ecological economics, which challenges the foundational assumption of unlimited economic growth. As the authors of *The Limits to Growth* (Meadows et al., 1972) underline, every economic system operates within the biogeophysical boundaries of the planet, and exceeding them leads to irreversible environmental consequences. In this spirit, ecological economics calls for abandoning the paradigm of accumulation in favor of ecological balance and long-term sustainability—the core principles of sustainable development.

As Laville (2010) observes, the solidarity economy seeks to re-embed economic activity in social and ecological relations, meaning that market mechanisms should be subordinated to community-oriented values rather than the other way around.

In this context, CSA gains new significance as an example of an economic practice based on cooperation, care, and the common good. CSA can be interpreted as part of the broader current of the solidarity economy, which asserts that the economy should serve people and the environment—not the reverse. Such initiatives align with the call to rebuild local systems of production and consumption that strengthen trust, responsibility, and intergenerational justice.

In summary, critical approaches to the neoliberal development paradigm not only expose its limitations but also propose alternative visions of the economy—more equitable, shared, and socially rooted. Models such as CSA offer concrete examples of these ideas in practice: they reject the logic of profit maximization in favor of a relational and sustainable understanding of development, where the central categories are responsibility and community rather than growth. In doing so, they form part of a broader transformation toward an ethical and ecological economy that emphasizes the interdependence between people, society, and the environment.

2.3 The Theory of Sustainable Development and Post-Growth

Economists critical of the neoliberal paradigm argue that economic growth cannot continue indefinitely due to the finite supply of natural resources and the inherent flaws of the market. This idea is succinctly expressed by S. Latouche (2010), who stated: "dirty, neoclassical growth cannot be infinite in a finite world." In practice, this is manifested in persistently high levels of environmental pollution and the overexploitation of both renewable and non-renewable natural resources.

A similar view was expressed by K. Polanyi (1944), who argued that neither nature nor people should be treated as commodities whose prices are determined solely by the market. According to Polanyi, laissez-faire economics was never a natural phenomenon, and the free market would never have emerged had events been left to unfold on their own. An alternative to the neoliberal development paradigm emerged in the form of the concept of sustainable development, reflecting society's growing awareness of the risks associated with continuous economic growth and finite natural resources. Although initially transformative in character, this concept gradually lost part of its critical potential. Consequently, post-growth and degrowth approaches have gained prominence in the literature, emphasizing the need to abandon GDP growth as the primary objective of social development (Jackson, 2009; Raworth, 2017; Kallis, 2018).

The term "sustainable development" was officially introduced at the United Nations Conference on the Human Environment in Stockholm in 1972. Since then, many definitions of the concept have been proposed, but most refer to the formulation presented in the Brundtland Commission's report Our Common Future (Brundtland, 1987). According to this definition, sustainable development is defined as "development that meets the needs of the present without compromising the ability of future generations to meet their own needs."

As noted, many authors have referred to this approach, including Pearce et al. (1990), who identified three core objectives of sustainable development: socio-economic well-being, intra- and intergenerational equity, and security.

Meanwhile, Rogall (2012), a representative of ecological economics, argues that sustainable development seeks to ensure adequate ecological, economic, and socio-cultural standards for all people living today and, in the future, within the limits of Earth's natural carrying capacity, while adhering to the principle of both intra- and intergenerational justice.

Based on a literature review, it can be assumed that the overarching goal of the sustainable development concept is the long-term improvement of the quality of life for present and future generations. This goal is pursued through the integration and balancing of five fundamental dimensions (Adamowicz and Dresler, 2006):

- Economic order, defined as the capacity for economic growth and increasing income levels;
- Ecological order, understood as the self-renewing ability of ecosystems;
- Social order, interpreted as the capacity of societies to improve living standards, foster development and self-realization, and nurture ethical values and awareness;
- Institutional order, shaping the social rules of conduct and the system of institutions (social, economic, and political);
- Spatial order, which integrates and organizes all forms of capital within physical space.

Given the specific characteristics of rural areas, it is particularly important to simultaneously incorporate economic, ecological, and social dimensions into agricultural activities.

According to Slangen (2001), sustainable development in agriculture involves three key dimensions. In economic terms, it means the capacity to meet the food needs of society; in social terms, the development of institutions (both formal and informal) that guarantee food security and nature protection for the entire population; and in ecological terms, the agricultural sector's ability to maintain natural resources in good condition. This perspective emphasizes a holistic approach that recognizes the interconnections between economic, social, and ecological aspects of the agricultural sector, and underscores the need to balance these three dimensions in order to achieve long-term stability and societal well-being.

Vereijken (1997) adds that meeting these conditions requires a stable production of adequate quantities of food of desired quality, at prices acceptable to society. At the same time, farmers should receive incomes comparable to those of other professional groups, with efforts to maintain the highest possible level of agricultural employment. This must be accompanied by the protection of soil, water, air, the agricultural landscape, and the welfare of farm animals.

In conclusion, sustainable development can be understood as the coherent pursuit of economic, environmental, and social objectives. According to Fiedor (2002, pp. 225–255), this requires a re-evaluation of the traditional understanding of the relationship between humans (society), the natural environment, and the economy.

Important reflections on sustainable development were offered by Kołodko (2014), who emphasized the importance of triply sustainable, long-term socio-economic development. This triple sustainability refers to:

- Economically sustainable growth, in terms of goods and capital markets, investments, finance, and the labor force;
- Socially sustainable growth, in terms of fair and socially accepted income distribution, equitable access to public services, and labor market participation for key population groups;
- Ecologically sustainable growth, in terms of maintaining a proper balance between economic activity and nature.

Thus, Kołodko highlights the need for internal balance within each of these domains.

Expanding on these ideas, one can argue that sustainable development requires two fundamental conditions:

- External balance: equilibrium among the various dimensions (economic, ecological, social, institutional, and spatial);
- Internal balance: harmony within each of these dimensions.

In summary, sustainable development should be characterized by durability and self-sustainability. Self-sustainability refers to the creation of reserves and incentives for further development, while durability implies that none of the elements of development should be diminished in the process (Kociszewski, 2013, p. 29).

In the context of rural areas, sustainable development also involves the harmonious achievement of multiple objectives. Woś and Zegar (2002, p. 51) identify four priorities in their concept of sustainable rural development:

- The protection of rural areas, including the preservation of rural landscapes, biodiversity, and the prevention of soil erosion;
- The protection of water, soil, and air from agricultural pollution;
- A cautious approach to the development of biotechnology and genetic engineering;
- The stable and sustainable development of agriculture.

From this perspective, CSA can be seen as a practical example of sustainable development in action. CSA integrates the economic, social, and environmental pillars of sustainability while maintaining spatial order typical of rural areas.

In economic terms, CSA increases the income stability of farmers and guarantees financial liquidity by eliminating intermediaries in the supply chain. A key issue in analyzing CSA is price. It should be noted that the initial purchase price of food delivered directly from the farmer tends to be higher than that offered by retail chains. However, due to cost reductions—especially those resulting from shorter supply chains—the final product price may be lower.

Higher prices for CSA products also compensate farmers for the increased labor inputs required for their production. Moreover, CSA is inherently based on traditional production methods, which reduces the demand for capital-intensive technologies. It can therefore be argued that the CSA model helps to restore the balance between labor and capital.

In social terms, the implementation of CSA principles may help to reduce rural depopulation. This can be attributed to increased farmer incomes (due to higher product prices), combined with the maintenance of existing agrarian structures.

CSA also positively influences the environmental dimension by reducing the use of synthetic fertilizers and pesticides. Additionally, the model supports biodiversity preservation, unlike many large-scale farms which often operate monocultures, such as oilseed crops. The preservation of small farms may also mitigate the negative phenomenon of farmland being converted into residential areas—a process associated with urban sprawl that leads to irreversible changes in rural spatial structure. As a result, rural areas lose their distinctive character and often become "bedroom communities" for urban centers.

Unfortunately, as statistics show, the overall progress of economies toward sustainable development is occurring, as Jackson (2009) puts it, "depressingly slowly" and is increasingly a manifestation of political rhetoric rather than genuine action. Consequently, contemporary literature distinguishes two main approaches:

- Green growth, which assumes the possibility of *decoupling* economic growth from resource use through technological innovation and improved energy efficiency (OECD, 2011);
- Post-growth/degrowth, represented by authors such as Jackson (2009), Latouche (2010), Raworth (2017), Daly (1996), and Kallis (2018), who argue that such decoupling is largely illusory and insufficient to prevent environmental degradation (Fletcher and Rammelt, 2017).

Post-growth perspectives call for an economy operating within the ecological limits of the planet and grounded in the values of solidarity, cooperation, and quality of life rather than unlimited consumption. As T. Jackson (2009) notes, prosperity in post-growth societies should be measured not by the rate of production growth but by levels of social trust, health, leisure time, and interpersonal relationships.

Essentially, the post-growth concept rests on three assumptions (Koch and Buch-Hansen, 2020):

1. A substantial share of ecosystem (planetary) resources is inherently finite;
2. Unlimited demographic and economic growth in a world of limited resources is impossible;
3. GDP is not an adequate measure of economic progress.

A common feature of this approach is the association of post-growth with the "deconstruction of reality" or the "detoxification of the environment." Its innovative character lies in creating space for new forms of creativity and innovation: it offers a framework for exploring alternative development pathways by setting different economic goals and recognizing new ecological and social boundaries (Matysiak and Raftowicz-Filipkiewicz, 2017).

Post-growth concepts differ from other economic categories in that they do not imply negative or zero growth achieved through restrictions, sacrifices, or production cuts. Growth remains important because it enables long-term development; however, in this perspective it refers to qualitative rather than quantitative forms of growth—"clean" and "green"—which are not measured by GDP indicators (Khmara and Kronenberg, 2020; Raftowicz, 2023).

In this context, K. Raworth (2017) introduces the metaphor of "doughnut economics," according to which development must occur within a safe and just space bounded by a social foundation (ensuring the satisfaction of basic human needs) and an ecological ceiling (the planetary boundaries). Crossing either boundary leads to social inequality or environmental degradation. G. Kallis (2018) adds that degrowth does not signify recession but a deliberate transformation toward ecological justice and community well-being.

In response to the limitations of macro-scale sustainable development strategies, micro-social approaches have gained increasing importance, highlighting the role of everyday practices and grassroots initiatives. Concepts such as "quiet sustainability" (Smith and Jehlička, 2013) and "everyday resilience" (Adger, 2000; Scoones, 2016) emphasize that transitions toward sustainable living occur not only through public policies but also through daily actions—community gardens, local exchange networks, consumer cooperatives, or CSA initiatives. Such micro-social innovations build social resilience and strengthen communities' capacity for self-regulation and cooperation.

From this perspective, CSA represents a practical expression of post-growth principles. This model is based on cooperation, solidarity, and trust-based relationships between farmers and consumers, rejecting the logic of profit maximization in favor of responsibility, reciprocity, and local well-being.

2.4 The Theory of Social Capital

In the literature, social capital has been defined in numerous ways. Scholars approach the concept from different angles, analyzing it at various levels and emphasizing different elements—sometimes treating it as a private good, and at other times as a public good. Researchers typically focus on three fundamental aspects: the structural (i.e., the structure of interpersonal connections), the normative (i.e., social norms relating to cooperation, patterns of value fulfillment and interests, trust, solidarity, and custom), and the behavioral (i.e., specific manifestations of cooperation, assistance, and information exchange). Among the classic definitions of social capital are those proposed by Bourdieu (1986), Coleman (1990), Fukuyama (1995), and Putnam (1993).

Bourdieu (1986) defines social capital as the sum of actual and potential resources linked to possession of a durable network of more or less institutionalized relationships based on mutual acquaintance and recognition. For Coleman (1990), social capital is defined by its function. It consists of certain aspects of the social structure that facilitate actions of individuals within that structure. Social capital is productive and enables the achievement of goals that would be unattainable in its absence. Fukuyama (1995) views social capital as an instantiated informal norm that promotes cooperation between individuals. Putnam (1993), meanwhile, understands social capital as referring to features of social life such as trust, norms, and networks that can improve the efficiency of society by facilitating coordinated actions.

To summarize, social capital determines the level of trust among economic actors, which is the foundation for all economic exchange and cooperation. Market competition, by definition, assumes a "game" governed by established "rules of play." These rules depend on social capital—on institutions and the credibility of the players. This means that social capital is essential for the "reproduction of capital," whether it be natural, physical, human, or even financial.

In this framework, social capital plays a superior role in the economic system (Matysiak and Raftowicz, 2019), and consequently, in socio-economic development (Serageldin and Grootaert, 2000). Social capital includes the characteristics of social organizations such as networks of individuals or households and the associated norms and values that generate externalities benefiting the broader community. Social capital is shaped by three main determinants (Matysiak, 1999, pp. 61–62):

- Social trust (Putnam, 2000);
- Institutions regulating interactions between people and their access to resources, consisting of formal constraints (e.g., rules, laws, constitutions), informal constraints (e.g., norms of behavior, conventions, self-imposed codes), and specific methods of enforcement (North, 1994);
- Norms of reciprocity and the credibility of actors (Six et al., 2015).

The most important function of social capital is the coordination of individual and group activities within the economy. The higher the level of social capital, the easier it is to form new economic organizations. The most efficient organizations are created within communities that share the same ethical norms (Fukuyama, 1995). Social capital fosters the growth of social welfare by enhancing people's capacity for spontaneous action. Social trust increases the economy's organizational adaptability to changing technological conditions in production, transportation, and consumption. When people trust each other, they are not afraid of organizational innovation or structural reform (Matysiak, 1999, pp. 61–62).

From a theoretical perspective, there is a strong link between social capital and the functioning of the CSA model. On one hand, a high level of social capital facilitates the formation of such cooperation; on the other hand, its absence poses a barrier to CSA development. Conversely, CSA helps build social capital by fostering relationships between farmers (producers) and consumers.

The theory of social capital can also be useful in explaining differences in the significance of CSA across countries. Moreover, it may provide a framework for domestic research, especially in countries like Poland, where different regions have developed under distinct institutional conditions.

When examining the CSA model through the lens of social capital, it is essential to consider:

- Social trust: farmers are more willing to cooperate when there is a high level of trust between them and consumers (Zoll et al., 2023),
- Norms of cooperation: CSA is based on reciprocity and a shared sense of responsibility for outcomes,
- Social institutions: local organizational structures and formal agreements enhance the durability of relationships among participants.

Based on these assumptions, one can infer that CSA is more likely to flourish under conditions of so-called long duration, accompanied by gradual, evolutionary development. As a result, the population becomes deeply rooted and shares similar values and norms, which helps to foster trust.

Unfortunately, in Central and Eastern Europe, these conditions are generally not met, which may explain the difficulties in implementing this model of cooperation. This draws attention to the relationship between social capital and human capital. In post-communist countries, a high level of human capital seems to facilitate the development of social capital and may serve as an alternative to "long duration." It may even be argued that human capital is a necessary condition for building social capital. This is due to the region's historical context—post–Second World War border changes and accelerated industrialization led not only to the construction of large factories but also housing estates, triggering population migration and the breakdown of established social ties. These factors were not conducive to building trust.

In such cases, the role of a well-educated and prepared local leader becomes particularly important. Such individuals can help overcome mental barriers and act as initiators of cooperation at the local level.

In the Polish context, the issue of social capital acquires particular significance. Research conducted by the Polish Economic Institute in 2024 (PIE, 2024) shows that despite ongoing institutional and economic development, levels of interpersonal trust in Poland remain low, hindering the emergence of cooperative networks and community-based initiatives.

These findings confirm earlier analyses of the weakness of civil society in post-communist Europe, which—as noted by Howard (2003) and Sztompka (1999)—point to persistent patterns of distrust and social atomization. According to Howard, the fragility of civic engagement in the region stems from the legacy of state socialism, which devalued the role of voluntary associations and collective action. Sztompka further emphasizes that entrenched mistrust, atomized social relations, and limited capacity for collaboration continue to pose barriers to grassroots mobilization.

At the same time, recent years have seen the development of alternative forms of cooperation: informal producer–consumer networks, cooperative groups, and CSA initiatives, which represent examples of bottom-up reconstruction of social capital. The activities of the EkoRozwój Foundation (FER) and other organizations promoting food sovereignty illustrate how education and community practices can strengthen trust and enhance participants' sense of agency.

According to the SolAbility Global Sustainable Competitiveness Index (SolAbility, 2024), Poland—like most post-communist countries—scores relatively low in the category of social capital compared with Western European states, confirming the persistence of institutional and social gaps. Nonetheless, these reports also highlight the developmental potential of local initiatives, which may serve as laboratories for rebuilding trust and cooperation based on solidarity and the common good.

Against this backdrop, CSA may thus be understood as a form of socio-organizational innovation that not only supports local food systems but also reinforces interpersonal relations and trust—key elements for the long-term resilience and development of local communities.

2.5 The Theory of Innovation

There is no universally accepted definition of innovation in the academic literature. Both theorists and practitioners assign different meanings to the concept of innovation. Joseph Schumpeter (1934) is widely regarded as a pioneer in innovation research, having been the first to introduce the concept into academic discourse. In Schumpeter's view, innovation occurs in five specific cases: (1) the introduction of a new product or a new type of product, (2) the introduction of a new method of production, (3) the opening of a new market, (4) the acquisition of a new source of raw materials or semi-finished goods, and (5) the implementation of a new organization of an industry. Thus, innovation in Schumpeter's framework encompasses all activities related to development, change, and improvement.

A different approach is presented by Drucker (2006), who defines innovation as a specific tool of entrepreneurship—an activity that endows resources with new capacities to create wealth. This definition highlights that innovation is not limited to technological breakthroughs or short-term trends, but refers to strategic actions that enable the creation of new value.

Numerous other definitions can also be found in the literature, ranging from narrower interpretations, such as Freeman's (1982) view of innovation as the first commercial application of a new product, process, or device, or Mansfield's (1968) definition, which sees innovation as the first use of an invention, to broader perspectives. For instance, Kotler (2017) argues that innovation refers to any good, service, or idea that is perceived as new by someone.

Parker (1974) considered innovation to be a process involving activities that bring a new product or production method into practical use. Similarly, Harman (1971) defined innovation as the introduction of new or significantly improved products or processes into the market.

The evolution of the concept of innovation is also evident in successive editions of the Oslo Manual (1997), where innovation is defined as the implementation of a new or significantly improved product (good or service), process, marketing method, or organizational method in business practices, workplace organization, or external relations.

In the context of CSA, the contributions of Silverberg and Soete (1994) and Fagerberg (2005) are also noteworthy. The former emphasized the importance of human capital in driving innovation, while Fagerberg viewed innovation as new and improved solutions—compared to those previously in use—that influence socio-economic living conditions. It is important to note that one of the objectives of CSA is to improve the living standards of both farmers (by ensuring stable income) and consumers (by providing access to high-quality food products). At the same time, participation in CSA requires a high level of awareness and expertise among its members—factors closely tied to a strong foundation of human capital.

To summarize, when adopting a broad view of innovation, it can be argued that innovation does not have to be objectively novel; rather, what matters is how it is perceived by the market or society. This means that innovation may consist not only of entirely new inventions, but also of new applications, improvements to existing products, or innovative approaches to problem-solving that create additional value and are perceived as new by their audience.

Innovation is not limited to the creation of new products. It also includes improvements in existing production processes, which may lead to greater efficiency or cost reduction. Moreover, innovation can involve marketing methods, such as new ways of promoting products, entering new markets, or transforming communication strategies with customers. Organizational innovations, in turn, refer to the implementation of new management or work organization methods that can improve company performance—for example, through better utilization of human resources or more efficient supply chain management. Importantly, innovation may also involve establishing new relationships with external stakeholders, such as strategic partnerships that help companies better adapt to changing market conditions (Oslo Manual, 1997).

In recent years, the literature has increasingly emphasized the importance of socio-technical innovations, which integrate technological, social, and organizational dimensions within processes of systemic change. Unlike classical technological innovations, which focus on a specific product or process, socio-technical innovations involve the transformation of entire socio-economic systems, combining technological progress with new models of cooperation, governance, and social relations (Geels, 2005; Smith et al., 2010; Seyfang and Longhurst, 2016). Their essence lies in the co-creation of solutions by diverse groups of stakeholders—producers, consumers, public institutions, civil society organizations, and researchers—based on the values of the common good and long-term sustainability. In this sense, socio-technical innovations form a foundation for food system transitions, leading to enhanced resilience, greater equity, and stronger local embeddedness.

In the case of rural areas, it is impossible to discuss innovation and innovativeness without taking into account their specific characteristics. The unique nature of these areas means that innovativeness is not an end in itself, nor should it serve solely to increase the productivity of productive resources. Innovation should support sustainable development. A compelling argument can be made that innovation ought to be subordinate to the goals of sustainability.

Therefore, when analyzing the issue of innovativeness in rural areas, it is essential to distinguish between constructive innovations—those aligned with the paradigm of sustainable development and destructive innovations—those inconsistent with the paradigm of sustainable development (Struś, 2018).

Currently, it appears that destructive innovations dominate rural areas. The source of these innovations is not farmers themselves, but external actors. These actors expect agriculture to meet the demands of mass consumption, which necessitates increased productivity. They exert pressure on farmers to deliver large volumes of standardized products. Such expectations encourage productivity-oriented innovations that do not contribute to sustainable development.

The continuation of this policy may paradoxically result in a situation where, in the long run, innovation becomes not a driver of progress but a source of stagnation and regression in rural areas. It is important to emphasize that innovations not aligned with sustainable development risk generating private gains while imposing social costs (Rodrik, 2015).

To summarize, innovativeness in rural areas should not be measured primarily by the adoption of new technical or technological solutions, but rather by organizational and social transformation. The goal of innovation should be to stabilize the market and establish direct relationships between farmers and consumers.

The CSA model serves as a prime example of constructive innovation, introducing novel solutions into the agricultural sector, such as:

- Product innovations: CSA promotes unique, high-quality products that stand out in the marketplace.
- Organizational innovations: the model eliminates intermediaries and creates new distribution channels based on direct relationships.
- Social innovations: the sharing of risk and benefits between farmers and consumers represents a novel approach to organizing production and consumption.

In conclusion, CSA represents an innovative approach to agricultural practice by directly connecting farmers with consumers, who become jointly responsible for the development of rural areas. At the same time,

CSA fulfills the criteria of constructive innovation that supports sustainable development. Unfortunately, this type of innovation is questioned by proponents of radical change in Polish agriculture. They seek to enhance the narrowly defined competitiveness of the sector, primarily through land consolidation and increased financial investment, coupled with labor reduction. While this approach aligns with the neoliberal development paradigm, it runs counter to the principles of sustainability.

3 Opportunities and Constraints for the Development of CSA in Poland in the Context of Agricultural and Rural Transformation

3.1 Agriculture and Rural Development under the Socialist Economic System

After the Second World War, Poland fell within the sphere of influence of the Union of Soviet Socialist Republics (USSR). As a consequence, a systemic transformation took place, leading to the establishment of a totalitarian state that controlled citizens' activities in the political, social, and economic spheres. In the economic domain, the transformation aimed to replace the capitalist economy with a socialist one—that is, a centrally planned command economy.

The new paradigm of economic development was modeled after the Soviet system. It was based on the assumption that planned rationality was superior to market mechanisms. The market was deemed incapable of ensuring optimal development; therefore, this responsibility was assigned to the state. As a result, efforts were made to eliminate private ownership and the market itself through the nationalization of the means of production. At the macroeconomic level, this system resulted in distorted cost, price, and exchange structures; enforced negative structural adjustments across sectors and regions; immobility of resources and production factors; and a decline in initiative and entrepreneurial spirit (Lodkowska-Skoneczna et al., 1996).

Ultimately, this gave rise to an economic system characterized by:

- The dominance of state ownership of the means of production;
- The primacy of industries producing capital goods;
- Underinvestment in agriculture and consumer goods industries;
- Limited consumption;
- Central planning and management;
- An extensive state and economic administrative apparatus ensuring hierarchical coordination and control of activities;
- Strong involvement of the ruling party in production and economic decision-making.

DOI: 10.4324/9781003745198-4

The above-mentioned principles of the socialist economy were also applied to agriculture. This involved the establishment of State Agricultural Farms (Państwowe Gospodarstwa Rolne, or PGRs) and production cooperatives, along with repressive measures targeting peasant farms.

The starting point for changes in agriculture was land reform. According to the declarations issued in the Manifesto of the Polish Committee of National Liberation (PKWN) on 22 July 1944, land intended for resettlement and the creation of new farms was to be sourced from the liquidation of large landed estates (over 50 ha, or over 100 ha in the Poznań, Gdańsk, and Silesian voivodeships) as well as from properties confiscated from Germans and individuals deemed traitors to the Polish nation (Manifest PKWN, 1986). The detailed provisions of the reform were set out in September 1944 (*Journal of Laws*, 1944), and for the western and northern territories, not until September 1946 (Dekret, 1946).

The course of the reform varied across regions. In the most overpopulated voivodeships, the supply of land obtained through the liquidation of large estates was quickly depleted (in the Rzeszów and Lublin regions, for example, land sufficed for only about 20% of applicants). The standard allotment of 5 ha for newly established farms thus proved unsustainable. In practice, the average size of new farms in the eastern territories was approximately 2.8 ha (Kaliński, 2005, p. 10).

The situation was different in the former Prussian partition, where relatively large peasant farms had already existed before the Second World War. The 5-ha standard for newly established farms was therefore unattractive—especially since there was a substantial amount of land available for redistribution, including former German properties designated for parceling. It was only after the standard was raised to 7 ha that peasant interest in land redistribution temporarily increased. In practice, however, newly created farms in Wielkopolska were often even larger, ranging from 10 to 12 ha. It was only after subsequent state inspections that these holdings were reduced to the officially prescribed 7 ha.

An entirely different situation unfolded in the western and northern territories incorporated into Poland after the Second World War, where all land had been seized by the state as former German property. These territories occupied a key position in the plans of the new authorities. They were intended to accommodate the rural population from the former eastern territories—who had not received land during the initial reform—as well as Poles repatriated from Soviet territory. A significant land surplus and difficulties with settlement made it possible to apply higher land allotment norms than in other regions—ranging from 7 to 15 ha.

By 1949, approximately 6 million ha of land had been redistributed, including over 3 million ha in the western and northern territories. Around 13,000 landed estates were either parceled or taken over by the state (Batowski, 2009, p. 149). As a result, in the prewar Polish territories,

347,000 new farms were created and 254,000 existing ones expanded. The average size of newly established farms was 5.4 ha. In the newly incorporated territories, 467,000 farms were established, with an average size of 7.9 ha (Kaliński and Landau, 2003, p. 203).

Small farms were economically weak, unable to support intensive production, and highly susceptible to tax pressure. As a result, they were more vulnerable to state efforts aimed at the collectivization of agriculture. At the same time, some landed estates were excluded from redistribution to serve as the foundation of a state-run agricultural sector. In total, over 3 million ha remained in state hands, with the majority (around two-thirds) located in the western and northern territories.

At the turn of the 1940s and 1950s, the authorities undertook efforts to transform Polish agriculture along Soviet lines. This was to be achieved through collectivization, the reduction of individual (especially large) farms, and the expansion of the state sector. Despite a propaganda campaign in 1949—the first year of collectivization—only 243 cooperatives were formed, representing just 0.1% of all farms. This low figure led the authorities to adopt administrative measures, political repression, physical coercion, and economic sanctions to force peasants into joining cooperatives. Resistant farmers were subjected to increased tax burdens, grain and livestock requisitions, fines, arrests, and even deportation to labor camps. The countryside was also burdened with additional levies, such as a mandatory contribution to the Social Fund for Agricultural Development.

The pace of cooperative formation varied. It was generally slower in the prewar Polish territories, while in the western and northern territories—where resettled and redistributed farms dominated—it was faster. By the end of 1955, nearly 10,000 cooperatives had been established, managing approximately 1.9 million ha of farmland, or around 9% of the country's agricultural land area (Kaliński and Landau, 2003, p. 249).

Although individual farms were initially declared the foundation of the agrarian system in postwar Poland, in practice, the creation of a state agricultural sector began during the very implementation of land reform, with 10% of the total agricultural land area allocated for this purpose. In 1946, the State Land Properties Enterprise (*Państwowe Nieruchomości Ziemskie*, PNZ) was established to take over and manage abandoned and devastated estates in the western and northern territories. PNZ's operations were soon extended to the prewar Polish territories as well, although the state acquired significantly less land there (Rogala, 1972).

At the beginning of 1949, all state-owned farms were consolidated into State Agricultural Farms. These were intended to serve as model farms that would ensure a stable food supply for the urban population, which was growing as a result of industrialization. However, despite preferential treatment, the efficiency of the PGRs was significantly lower than that of individual farms. In 1955, gross production per hectare of agricultural land amounted to 621 złoty in peasant farms, 517 złoty in production cooperatives, and only 394 złoty in PGRs. Grain production alone, thanks to mechanization, reached levels comparable to those of peasant farms (Skodlarski, 2007, p. 404).

By 1955, over 6,000 state-owned farms had been established, covering approximately 12% of agricultural land. More than 60% of PGRs were located in the western and northern territories, and around 25% in the former Prussian partition areas. The remaining were distributed across other parts of the territory of the Second Polish Republic. Over the following years, the area of land under state control continued to expand, reaching its peak in 1990. The highest concentration of PGR-managed land—over 50% of total agricultural area—was in the Koszalińskie and Szczecińskie voivodeships.

After 1956, significant changes occurred in Polish agriculture. Most importantly, the majority of production cooperatives disbanded. Of the nearly 10,000 cooperatives that had been established, only around 1,500 remained. At the same time, the state's policy toward individual farmers shifted. Obligatory deliveries of agricultural produce were reduced, procurement prices were increased, tax policies were relaxed, and farmers were granted access to bank credit. As a result, agricultural output increased.

Further changes in state policy toward individual farmers occurred in the 1970s, prompted by the country's deteriorating food supply situation. In 1972, obligatory deliveries were abolished, and farmers gained access to investment credit. Although state-owned farms remained a priority in the distribution of agricultural inputs, individual farmers could, in return for contractual agreements, receive allocations of scarce machinery, tools, and fertilizers (Mędrzecki, 2003, p. 156). Farmers were also permitted to expand their holdings to 8 ha, and in some cases up to 20 ha. At the same time, the division of land into farms smaller than 8 ha was prohibited to prevent excessive fragmentation. For the first time, farmers and their families were granted access to free healthcare.

After 1980, the legal status of individual farms improved (e.g., restrictions on land inheritance were lifted). However, the economic conditions in agriculture remained difficult. Low levels of mechanization, insufficient domestic production of pesticides and fertilizers, and limited access to agricultural inputs meant that output remained far below Western European levels. Food shortages, already severe, became more pronounced. The situation worsened significantly after the introduction of martial law, when Western sanctions led to a suspension of imports—including animal feed—which triggered a collapse in livestock production.

To summarize, during the era of the Polish People's Republic (*Polska Rzeczpospolita Ludowa*), four main forms of land ownership existed:

- Individual farms
- State Agricultural Farms
- Production cooperatives
- Agricultural circles (*kółka rolnicze*) with their own land use structures (Gorzelak, 2010)

The state's approach to agriculture evolved throughout the socialist period. Although state and cooperative farms were consistently favored, apart from the years 1948–1956, the authorities did not pursue the forced nationalization of private farms. On the other hand, financial restrictions prevented individual farms from developing effectively. This contributed to the entrenchment of a highly diversified agricultural structure across different regions of Poland. In the territories of former Poland, small individual farms ranging from 0.5 to 5 ha, oriented toward subsistence agriculture, predominated. In contrast, in the western and northern territories, medium-sized farms of 7–15 ha were more common and had a significantly more market-oriented character (Kociszewski, 1999, p. 223). The agrarian structure shaped during this period had a profound influence on the transformations in Polish agriculture that followed after 1989.

3.2 Transformations in Rural Areas after 1989

The systemic transformation initiated in Poland in 1989 led to structural changes in Polish agriculture and influenced its subsequent development. Following the marketization of the economy and, later, Poland's accession to the European Union, a significant accumulation of capital occurred, contributing to changes in the agrarian structure. The processes of agricultural concentration intensified—particularly in terms of land ownership and production, but also in capital directed toward on-farm investments.

These processes were supported by the adoption of a neoliberal paradigm of economic development, the intensification of globalization, and the acceleration of technological progress. In particular, globalization resulted in the growing influence of large corporations in the agri-food processing and retail sectors. These corporations favored large-scale farms due to product quality, delivery timeliness, batch size, and lower transaction costs (Zegar, 2009).

The emergence of new capital-intensive technologies and advances in biotechnology also supported the concentration process. In both of these cases, large farms were the primary beneficiaries of the changes, which weakened the competitive position of smaller-scale farms (Zegar, 2009).

At the same time, increasing competition on European and global agricultural markets became evident. This competition forced farmers to take measures aimed at improving the economic efficiency of production and exploiting economies of scale to reduce unit costs of agricultural goods (Rynio, 2013).

In agriculture, concentration was also facilitated by imperfect market structures (le Vay, 1983). Agricultural markets are characterized by asymmetry: a high degree of concentration among suppliers of production inputs and buyers of agricultural products, and a low level of concentration among agricultural producers. A way to reduce this imbalance in market power was through the expansion of land holdings and both horizontal and vertical integration (Małysz, 2002).

In Poland, the roots of agricultural concentration can be traced back to the years 1992–1998, when State Agricultural Farms (*Państwowe Gospodarstwa Rolne*, PGRs) were liquidated. As a result, approximately 4.5 million ha of land were redistributed. The spatial distribution of this land was highly uneven, with the highest shares of former state-owned land located in the present-day Zachodniopomorskie, Warmińsko-Mazurskie, Pomorskie, and Dolnośląskie voivodeships.

According to the 1991 Act on the Management of Agricultural Property of the State Treasury, which established the Agricultural Property Agency of the State Treasury (*Agencja Własności Rolnej Skarbu Państwa*), the early years of privatization favored the sale and lease of large, often whole units. This led to strong land and capital concentration. Surrounding assets were often neglected, and animal husbandry was frequently abandoned in these large units.

The lack of policies favoring or strengthening family farms also resulted in the emergence of large landholdings, often acquired at prices equivalent to the value of a single year's commercial grain production per hectare of arable land. This led to a situation in which "the big got bigger." Medium and small farms generally did not benefit from access to state-owned land.

These developments contributed to a steady concentration of land ownership in Poland, accompanied by a decline in the total number of farms—especially small farms under 10 ha.

Although these processes have developed relatively slowly (Zegar, 2009), they are becoming increasingly evident. This is confirmed by the results of the 2010 and 2020 Agricultural Censuses (*Powszechne Spisy Rolne*). According to census data, the number of farms in Poland decreased by approximately 12.7% between 2010 and 2020—from 1.509 million to 1.317 million. This decline can primarily be attributed to a significant reduction in the number of small farms with an area of up to 10 ha. During the analyzed period, the number of farms in this size category fell from 1.160 million to 976,000—a decrease of about 15.9%.

From the perspective of the sustainable development paradigm, the decline in the number of small farms is a negative phenomenon. As earlier studies have shown, farms up to 10 ha in size are best positioned to implement the principles of sustainable development (Kutkowska, 2018). In theory, they also form the foundation for the functioning of Community-Supported Agriculture (CSA).

Between 2010 and 2020, the number of farms in the 10–30-ha range stabilized, while the number of farms exceeding 30 ha increased. Particularly notable was the growth in the number of farms classified as large (50–100 ha) and very large (over 100 ha), according to Polish standards. This trend reflects a shift in Polish agriculture toward an industrial model, which stands in contrast to the sustainable development paradigm strongly promoted by the European Union. It is thus possible to speak of the emergence of a dual model of agricultural development in Poland, where two parallel systems coexist: the industrial regime (industrial agriculture) and the sustainable regime (sustainable agriculture).

According to the Agricultural Censuses (*Powszechne Spisy Rolne*), between 2010 and 2020, the number of large farms (50–100 ha) increased by 52.9%, and the number of very large farms (over 100 ha) grew by 30%. As a result, in 2020, there were approximately 26,000 farms in the 50–100 ha category and 13,000 farms over 100 ha, compared to 17,000 farms and 10,000 farms, respectively, in 2010.

The increase in the number of large and very large farms, combined with the decline of small farms, meant that by 2020, farms in the 50–100 ha and over 100 ha categories accounted for approximately 2% and 1% of all farms, respectively. In 2010, these shares stood at 1.1% and 0.7%.

However, the progression of land ownership concentration is evidenced not only by changes in the number of farms. Equally important is the increase in the share of agricultural land held by the largest farms. In 2020, the 13,000 farms (1.0% of all farms) in the largest size category used 25% of the country's total agricultural area and 20% of the total sown area. Compared to 2010, this group of farms recorded a 3% increase in well-managed arable land and a 9% increase in sown area.

This growth, similar to that observed in other farm size categories, occurred at the expense of a reduction in other land types: permanent crops, fallow land, and permanent pastures. In the largest farm category, the rate of decline in these land types was significantly higher than among farms overall and amounted to over 70%, approximately 65%, approximately 64%, and around 30%, respectively (Powszechny Spis Rolny, 2021).

The simultaneous increase in both the number of very large farms and the land resources at their disposal has highlighted land-use practices that run counter to the principles of sustainable development. There is a growing trend toward monoculture systems, to the detriment of more diverse and sustainable crop rotations. This is evidenced by the increase in land allocated to industrial crops, including sugar beets, rapeseed, and turnip rape. Between 2010 and 2020, the share of industrial crops in total sown area rose by 10.5%, from 1.171 million ha to 1.294 million ha. During the same period, the area devoted to sugar beet cultivation increased by 19.4% (from 206,000 ha to 246,000 ha), while the area under rapeseed and turnip rape rose by 3.8% (from 945,000 ha to 981,000 ha).

The emergence of Alternative Food Networks, of which CSA is a part, has been a response to the ongoing processes of land ownership concentration and the expansion of industrial agriculture. As outlined in Section 1.1, AFNs encompass diverse forms of direct farmer–consumer cooperation; however, their specific manifestations in Poland reflect the structural changes that have unfolded since 1989. Due to their largely informal and decentralized nature, the overall scale of AFNs is difficult to assess, yet food cooperatives remain one of the most visible expressions of this trend in the Polish context. Although cooperatives share historical roots with traditional forms of agricultural and consumer cooperativism, contemporary food cooperatives display a number of distinctive features, such as their bottom-up character, the absence of formal structures and the lack of official membership (Skorupska, 2024). These characteristics, combined with a general reluctance to formally register organizations, mean that cooperatives tend not to have a permanent character. As a result, it is challenging to determine their real number—both in terms of the number of cooperatives and the number of participants. Some formally exist but in practice do not operate or remain "dormant."

Zbyrowska (2022) identified 37 functioning food cooperatives in Poland. Śpiewak et al. (2022) estimated that around 50 such initiatives exist. They also refer to data from an informal umbrella organization for food cooperatives in Poland, according to which the number is approximately 24. Based on self-reported descriptions published online, Skorupska identified 25 active food cooperatives (Skorupska, 2024).

The functioning of food cooperatives in Poland can be traced back to 2010, when the Warsaw Food Cooperative was established by a group of activists (Bilewicz and Śpiewak, 2019). This initiative was characterized by a small operational scale and a flat, non-hierarchical structure. Tasks such as coordination, packing, and weighing products were carried out on a rotational basis, while decisions were made by consensus during meetings typically held once a month or after each collective purchase. Coordination challenges gradually contributed to the emergence of a second organizational model, in which a rotating board was introduced.

In summary, Polish agriculture is undergoing continuous transformation, and its dual nature is becoming increasingly evident. On the one hand, large and very large farms—characteristic of industrial agriculture—are gaining prominence. On the other hand, small farms focused on local markets are gradually disappearing. Nevertheless, their numbers remain relatively high, which provides an opportunity for the development of CSA.

3.3 Current State and Development Factors of CSA—A Theoretical Approach

It is currently difficult to determine precisely how many CSA initiatives operate in Poland. This situation stems from several factors. First, the CSA model is not legally regulated in Poland; therefore, no formal criteria exist to determine whether a given form of cooperation qualifies as CSA. As a result, there are initiatives that use the CSA label despite failing to meet the fundamental criteria defined in the literature. These are more accurately described as food cooperatives focused solely on the direct sale of agricultural products—for example, at local markets or bazaars. On the other hand, there are also organizations that could adopt the CSA label but choose not to. This may be due to the fact that the concept of CSA itself is not widely recognized in Poland, and consequently, neither is the term.

Second, there is no central register in Poland in which CSA initiatives are recorded. Nor is there any obligation to register them in existing economic or civil society databases maintained by courts, relevant ministries, or the National Agricultural Support Center (*Krajowy Ośrodek Wsparcia Rolnictwa*). Although it is impossible to determine the exact number of CSA operations in Poland, it can be stated that their presence is not widespread and falls far short of the situation observed in countries like France or Germany.

It is worth emphasizing that the establishment of CSA initiatives is inherently bottom-up, arising from the efforts of farmers or conscious consumers seeking local, often organic, high-quality products. There is no institutionalized support from either local or national authorities. This situation also contributes to the difficulties in estimating the actual scale of CSA initiatives in Poland. At the same time, it should be noted that EU funds available under rural development programs play a supplementary rather than decisive role. They do not constitute a strong incentive to establish new CSAs but may, at most, support entities that are already in operation.

Based on a review of the literature, it may be concluded that there are conditions in Poland conducive to the development of CSA. These stem from both supply-side factors (related to farmers) and demand-side factors (related to consumers).

To begin with, research to date suggests that the CSA model is particularly suited to small family farms (under 10 ha). In this regard, the structure of Polish agriculture still meets this condition. As previously mentioned, in 2020 there were approximately 976,000 small farms of up to 10 ha in Poland.

The literature increasingly highlights that small farms, despite their limited production scale, play a crucial role in maintaining sustainable and resilient food systems. Empirical studies confirm that, owing to their higher production diversification, reliance on local resources and strong social ties, small farms exhibit greater adaptive capacity in the face of climate change, market crises, and disruptions in supply chains (Altieri and Nicholls, 2017; Darnhofer et al., 2010). Comparative analyses and meta-analyses have demonstrated that organic farms and those diversified in terms of crop structure and production organization show higher biodiversity and lower susceptibility to yield variability (Tuck et al., 2014; Ponisio et al., 2015). These findings further indicate that the diversification of farming practices and strong local embeddedness can reduce systemic risks in agriculture and foster the development of socio-ecological resilience in rural areas.

Recent research by Chiffoleau et al. (2025) shows that short food supply chains constitute an element of an alternative economy grounded in specific principles of cooperation and exchange, linking economic objectives with social values. Drawing on data from French AMAP farms, the authors emphasize the importance of social embeddedness and the so-called economy of moderation, in which relationships, trust and reciprocity serve as key factors of durability and resilience in local food systems. In this perspective, SFSCs are not merely an alternative marketing channel but a form of social and economic innovation that supports the balancing of economic, ecological, and ethical goals.

According to FAO reports (2021), small family farms produce around one-third of the world's food, contributing to food security, the preservation of genetic and landscape diversity, and the development of local markets. Similarly, IPES-Food (2024) underscores the importance of territorial markets and short supply chains as mechanisms that strengthen food system resilience and bring production closer to consumption. In this view, small farms can be considered fundamental units of sustainable development, integrating environmental, economic, and social dimensions. As Chicoma (2024) notes, supporting the transformation of food systems requires recognizing the role of small farms as active agents of change rather than merely beneficiaries of public policies.

It should be strongly emphasized that CSA represents an opportunity for small farms to survive under the pressure of growing competition from large-scale agricultural operations. Without organizational change, small farms will not be able to compete—especially in terms of pricing—since they do not benefit from economies of scale. Moreover, small farms are characterized by low labor productivity, as confirmed by research conducted by A. Nowak (2022). According to the neoliberal paradigm of development, the productivity of production factors is expected to be maximized. From this neoclassical perspective, small farms are destined to be eliminated, and their populations to migrate to cities, seen as centers of economic growth. Such an approach, however, stands in contrast to the principles of sustainable development, which emphasize the importance of small farms as actors that strengthen socio-ecological resilience and territorial cohesion in rural areas.

These assertions are supported by existing research. In 2022, the average farm size in Poland, measured in Utilized Agricultural Area (UAA), was 19.3 ha. Labor productivity for such a farm amounted to 186.6 working hours per hectare. The EU average was 142.2 hours per hectare. For comparison, the same indicator stood at 54.2 hours per hectare in Germany and 37.2 hours per hectare in France (Nowak, 2022). For farms below 10 ha, labor input is even higher, which calls into question the market viability of such farms in their current form.

The literature emphasizes that traditional measures of economic productivity fail to capture the full range of environmental and social benefits generated by small farms, which produce significant public goods, including landscape conservation, biodiversity protection, and the strengthening of social ties (Altieri and Nicholls, 2017).

At the same time, Góral and Rembisz (2017) emphasize that improvements in labor productivity are driven primarily by better technical equipment and improved organization of work. Given that enhancing the technical equipment of small farms is not always feasible, a change in work organization—such as participation in CSA—remains a promising alternative.

Another disadvantage facing small farms is their inability to ensure large, regular, and standardized deliveries of products. This limits their potential for cooperation with retail chains or wholesalers. For these farms, the best opportunity lies in shortening supply chains and selling directly to consumers.

The second key factor influencing CSA development is the shift in consumer attitudes. Research shows that consumer interest in organic products is on the rise (Gradka et al., 2023, p. 5). In 2023, 33% of consumers regularly (at least once a month) purchased organic food. In addition to health concerns, purchasing decisions were also influenced by socio-ethical considerations, including environmental protection and animal welfare.

This indicates that consumers are increasingly aware not only of the health implications of proper nutrition, but also of environmental issues—one of the cornerstones of the CSA model. In addition, the last several years have seen a steady increase in the overall affluence of Polish society. This may help reduce one of the key barriers to CSA development—namely, the higher price of organic and traditional food products.

A further supportive factor for CSA may be the evolving direction of the European Union's agricultural policy, which is shifting away from maximizing production through intensive farming methods. Although previous policies succeeded in increasing food output, they also gave rise to economic, social, and environmental problems. Surplus food production became a burden for both taxpayers and consumers. Moreover, the widespread use of fertilizers and pesticides has had a detrimental effect on the natural environment, as well as on the quality and nutritional value of food products.

3.4 Barriers to the Development of CSA

Based on a review of the literature, it can be concluded that CSA may offer a response to the current challenges faced by small farms. The main factors negatively affecting their operations include:

- Relatively frequent income fluctuations among farmers;
- Limited ability of small farms to compete on price with large-scale farms;
- Underdeveloped or depreciated infrastructure;
- Agriculture's dependence on large corporations, which favors monoculture farming;
- Still low (though growing) consumer awareness of the relationship between food quality and health (e.g., diet-related diseases).

However, as previous research has shown, CSA initiatives in Poland still face a number of barriers. These are diverse in nature, ranging from economic issues to insufficient support from the public sector, and include challenges related to the low level of human and social capital. It is important to emphasize that such barriers affect both farmers and consumers.

Arguably the most critical barrier is the low level of trust, which results from the still insufficient level of social capital in Poland. This issue applies to both farmers and consumers. The problem of low social capital, particularly in rural areas, has been confirmed by earlier studies conducted by Łabędzki and Struś (2014) and Błażejowska and Gostomczyk (2018), and has also been highlighted by Onyszkiewicz and Sylla (2025).

From the perspective of CSA's future, social capital is of fundamental importance. Trust is the foundation of the prepayment system, which mitigates financial risk for farmers. One could argue that trust is a *sine qua non* condition for the functioning of this cooperation model—especially since prepayments are one of the key factors motivating farmers to participate in CSA (Kutkowska, 2017, p. 12). A lack of trust also leads to rising transaction costs associated with drafting detailed cooperation agreements and monitoring compliance with their provisions. This problem becomes more pronounced as the number of CSA members increases. It should be emphasized, however, that this is not a phenomenon specific to Poland or other post-socialist countries. Even in Western European states, the issue of increasing transaction costs related to supervising group members has been observed (Opitz et al., 2017; van Oers et al., 2018).

Another significant barrier is the high labor demands of traditional and organic farming methods. Existing studies indicate that farmers believe CSA participation requires excessive labor input (Kutkowska, 2018, p. 14). At the same time, farmers are often willing to forgo potentially higher incomes in favor of what is increasingly seen as a luxury good: free time. One possible solution to this issue could be greater involvement of consumers in fieldwork.

A recurring challenge in implementing CSA in Poland is the geographic distance between producers and their main markets. The greatest demand for organic and traditionally produced food comes from major urban centers. Distance from cities—and the associated transport costs—limits the viability of CSA in peripheral rural areas. Paradoxically, CSA initiatives may benefit from the otherwise negative phenomenon of urban sprawl. While this process reduces available agricultural land due to housing development, it also brings potential consumers closer to farmers who may be interested in joining CSA networks.

It is also worth noting that a barrier to CSA development lies in the attitudes of farmers themselves. As studies have shown, many farmers do not perceive CSA as an opportunity for developing their farms. They do not associate CSA participation with strengthening their competitive position relative to larger farms. Moreover, some farmers are satisfied with the current financial performance of their farms (Kutkowska, 2018, p. 11). This may be due to the emergence of a specific model of small farming based on two dominant factors: EU subsidies and off-farm employment. Subsidies provide a steady income, reducing the need to implement innovation—including organizational change. At the same time, many small farmers take advantage of the well-developed labor market by working outside agriculture. Combining farming with off-farm employment leads them to focus on crops that require relatively little labor input, such as cereals.

Supply-side barriers are mirrored by demand-side constraints. Among consumers, there is often a lack of understanding of the CSA model itself. Consumers frequently adopt a *homo economicus* perspective, seeking to maximize personal utility, which they associate not with supporting agriculture, but with purchasing "healthy food"—typically organic or traditionally produced products. As a result, they prefer to buy such products at farmers' markets or bazaars, without entering into contracts or making prepayments.

Many consumers also do not see the need to share financial risk with farmers and are skeptical of paying in advance without a guarantee of receiving the full quantity of products agreed upon. This attitude reflects a relatively low sense of co-responsibility for local development, including the functioning of local food producers. Consumers often do not identify with their local communities. This may be due to an insufficient level of human capital—which should not be equated solely with formal education.

Another obstacle to CSA development is the relatively high price of its products. Despite being sold directly, the prices of CSA products are often higher than those in retail chains. Additionally, consumers do not always appreciate the quality of such products, placing more value on visual appeal than on health benefits. It is also important to highlight a frequently overlooked factor: time. As mentioned earlier, time is a luxury good—for both farmers and consumers. Consumers are often unwilling to spend time engaging in farm work or collecting products directly from farmers. Instead, they opt for more convenient and less time-consuming solutions.

Finally, a major barrier to CSA development is the lack of support from public institutions—both governmental and local. CSA initiatives are not targeted by dedicated support programs. They do not benefit from organizational, legal, or financial assistance. Moreover, public institutions fail to promote the CSA concept itself, resulting in low public recognition and awareness of the model among both farmers and consumers.

Nevertheless, the broader literature shows that small farms—owing to their strong local embeddedness and the relational nature of their cooperation with consumers—constitute a key component of sustainable food systems, as confirmed by analyses conducted by FAO (2021) and IPES-Food (2024).

4 Community-Supported Agriculture in Poland—Empirical Research

4.1 Materials and Methods

The empirical research was conducted in Poland, the largest agricultural country among the new member states of the European Union. In terms of agricultural land area, Poland ranks third in the EU, surpassed only by France and Spain. The structure of agricultural holdings in Poland is notably fragmented, with as many as 74% of farms covering less than 10 ha, as shown in Table 4.1.

Additionally, Poland's unique historical background makes it a particularly compelling area of research. As a post-communist country that was partitioned for 123 years under three empires (Prussian, Austrian, and Russian), Poland exhibits diverse patterns of socio-economic development across its regions. The prolonged influence of different political and economic systems has led to the emergence of distinct informal institutions (Struś and Raftowicz, 2025), which may significantly affect the functioning and development of initiatives such as Community-Supported Agriculture (CSA). Despite growing interest in alternative food systems, the adaptation of the CSA model to the Polish context has not yet been comprehensively analyzed—highlighting both the importance and the necessity of the present research.

Table 4.1 Number and Share of Agricultural Holdings in Poland in 2020

	Number of Agricultural Holdings (in thousands)	*Share of Agricultural Holdings (%)*
Total	1317	100.0
Up to 10 ha	976	74.0
10–30 ha	285	19.6
30–50 ha	44	3.3
50–100 ha	26	2.0
Over 100 ha	13	1.0

Source: Author's own elaboration based on Agricultural Census (2021).

DOI: 10.4324/9781003745198-5

The empirical study was conducted in three areas: (I) among producers—CSA farms, (II) consumers, and (III) institutions within the agricultural business environment.

Ad. I. In the case of CSA producers, a survey was conducted between March and May 2024 across six of the ten currently operating farms. These farms were listed on the wspierajrolnictwo.pl website. Despite repeated efforts, no responses were obtained from the remaining four farms.

The surveyed farms were located in the following voivodeships: Łódź, Mazovia, Greater Poland, and West Pomerania. In each of these regions, the agricultural structure is dominated by small farms of up to 10 ha, as illustrated in Table 4.2.

Figure 4.1 shows six municipalities—1. Będzino, 2. Łobżenica, 3. Budzyń, 4. Słubice, 5. Jabłonna Lacka, and 6. Aleksandrów, highlighted on the map)—where the surveyed CSA farms are located. These are distributed across four voivodeships, indicated on the map, along with their respective capitals. The research conducted was qualitative in nature and based on the CAWI method (Computer-Assisted Web Interview), which involves collecting data via online questionnaires. This method was selected for its numerous advantages, including the ability to reach a broad group of respondents within a relatively short period and at reduced costs compared to traditional data collection approaches.

However, significant challenges arose during the research process due to socio-cultural factors. Polish agricultural producers often exhibit skepticism toward academic research and are reluctant to share information. This may stem from concerns about external institutional oversight or a lack of trust in the anonymity of surveys. As a result, obtaining reliable and detailed data proved difficult and required additional efforts—such as follow-up phone calls—to build trust and motivate respondents to participate.

The questionnaire used in the study consisted of 22 questions, both open- and closed-ended. For the closed-ended items, a Likert scale was employed, enabling the assessment of respondents' attitudes (beliefs and perceptions) toward the phenomena under analysis.

The open-ended questions focused on:

- The characteristics of the farm (area, number of consumers, employment),
- The benefits for farmers from participating in a CSA,
- The challenges associated with operating within the CSA model.

Table 4.2 Number of Agricultural Holdings in 2020 in the Surveyed Voivodeships

Voivodeship	*Number of Agricultural Holdings (in thousands)*						*Share of Agricultural Holdings (%)*				
	Total	*Up to 10 ha*	*10–30 ha*	*30–50 ha*	*50–100 ha*	*Above 100 ha*	*Up to 10 ha*	*10–30 ha*	*30–50 ha*	*50–100 ha*	*Above 100 ha*
łódzkie	117	90	23	2	1	0	76.8	17.7	2.1	0.9	0.3
mazowieckie	208	152	48	6	2	1	72.9	20.0	2.7	1.2	0.4
wielkopolskie	116	71	33	6	3	2	62.2	10.9	5.1	2.7	1.4
zachodniopomorskie	29	16	7	2	2	2	54.5	33.8	7.1	8.1	6.6

Source: Author's own elaboration based on Agricultural Census (2021).

Figure 4.1 The map of Poland with the locations of the six studied CSAs.

The closed-ended questions concentrated on:

- The motivations for participating in a CSA,
- The perceived need for consumer involvement in fieldwork,
- The development prospects of the surveyed farm,
- The barriers to the development of CSA in Poland,
- The expected forms of institutional support.

A clear limitation of the study was the small number of surveyed farms (six entities in six municipalities), all located within four voivodeships, which may not fully reflect the situation in other regions of Poland.

Ad II. In the case of consumers, it was not possible to conduct research among CSA members, as producers cited commercial confidentiality regarding their clients' data. Consequently, the survey was carried out among employees of higher education institutions in Poland, following a methodology used in previous, similar studies (Raftowicz-Filipkiewicz and Le Gallic, 2018; Raftowicz et al., 2019). This group was intentionally selected, based on the assumption that, compared to the general population, they possess greater knowledge of alternative food systems and therefore represent a potential group of consumers interested in participating in the CSA model. The reasoning adopted by the authors of the monograph to justify the selection of respondents also draws on studies conducted in Germany—a country in which CSA has been developing rapidly. German research clearly shows that typical CSA participants are well-educated, financially secure individuals with stable professional positions, living in urban areas, environmentally conscious, and critical of the current food system (Bîrhală and Möllers, 2014; Blättel-Mink et al., 2017; Wellner and Theuvsen, 2018). It may therefore be assumed that employees of Polish higher education institutions largely meet these criteria.

A total of 174 respondents participated in the study. The research was conducted in March 2025, with the questionnaire distributed electronically. The survey consisted of 21 questions. The survey questionnaire was divided into four sections:

I. Demographic data: five questions concerning age, gender, type of employment, place of residence, and monthly net income per household member.
II. Knowledge of CSA: five questions relating to familiarity with the CSA concept, sources of information, awareness of CSA initiatives in the local area, participation in CSA, and associations linked to this model of cooperation.
III. Opinions on CSA: four questions concerning the perceived advantages and disadvantages of CSA, as well as consumers' willingness to engage in on-farm work.
IV. Potential interest in CSA: seven questions addressing consumers' purchasing preferences and the factors that might encourage them to participate in a CSA scheme.

As in the survey conducted among farms, the questionnaire included both open-ended and closed-ended questions. A Likert scale was also used in the study.

It should be noted that demand-side research conducted among university employees may suffer from limited representativeness, as this is a specific group with higher ecological awareness and potentially different purchasing habits than the average consumer. As such, the results may not fully reflect actual interest in CSA among the broader population.

In addition, the number of responses obtained relative to the total population covered by the survey was too limited to allow for the appropriate use of quantitative methods in analyzing the results. For this reason, the study has a qualitative character. Nevertheless, it provides valuable insight into Polish society and its attitudes toward CSA.

Ad III. In the case of research involving Agricultural Support Institutions, structured face-to-face interviews were conducted with representatives of:

- Municipalities where CSA farms were located (six municipalities),
- Provincial Agricultural Advisory Centers operating in regions where CSA farms were active (four provincial centers),
- The Ministry of Agriculture and Rural Development in Warsaw,
- The National Agricultural Support Centre (KOWR) in Warsaw,
- The Agricultural Advisory Centre in Brwinów.

The interviews focused on assessing the functioning of the CSA system in the respective region, promotional efforts, and initiatives aimed at strengthening cooperation. Respondents were also asked which institutions currently offer the strongest support for CSA development, which should be more involved, and what types of support are needed to enable more effective CSA operations at the municipal or regional level. These interviews were conducted in July and August of 2023 and 2024.

The interview method was chosen as it allows for an in-depth examination of key aspects of CSA functioning, such as the dynamics of participant relationships, cooperation models, and the degree of engagement by supporting institutions. Direct interaction with respondents enabled the collection of detailed, context-specific information that would be difficult to obtain through questionnaires alone. Interviews also helped clarify responses, contributing to a more nuanced understanding of the complex relationships and mechanisms underlying CSA operations. An additional advantage of this method was the opportunity to capture subjective opinions and personal experiences, adding a deeper, qualitative dimension to the analysis.

A limitation of the study is that the interviews focused primarily on agricultural support institutions, which may have limited the analysis of relationships with the private sector, such as local retailers or non-governmental organizations.

4.2 CSA from the Farmers' Perspective—Motivations, Benefits, and Challenges

The study covered six farms operating within the CSA model, located in four voivodeships in Poland: mazowieckie, łódzkie, wielkopolskie, and zachodniopomorskie. All surveyed CSA farms specialized in vegetable cultivation; however, some also offered additional products such as fruits, eggs, preserves, groats, flour, dairy, and meat—indicating a diversified activity profile.

The total area of these farms amounted to 103 ha, with approximately 75% of this land held by the two largest farms, whose respective sizes were

41.0 and 36.0 ha. The remaining farms were significantly smaller, with areas ranging from 4.0 to 13.0 ha. According to the findings, however, the total area directly allocated to CSA cultivation was only 10.4 ha. This indicates that, on average, each farm dedicated around 1.7 ha to CSA-related activities, although the cultivated area varied considerably—from as little as 0.2 ha to as much as 4.0 ha.

These proportions may reflect differences in operational models, production strategies, and the degree of engagement in the CSA system.

The research revealed that only two of the surveyed farms relied exclusively on CSA as their sole source of income (Table 4.3). This leads to the conclusion that the CSA model in Poland tends to represent an expansion of existing agricultural activities rather than a standalone business model. This finding challenges a frequently cited assumption in the literature that CSA initiatives are primarily undertaken by small farms (up to 10 ha). Consequently, it raises the question of whether the fragmented agrarian structure typical of Poland truly supports the emergence of this type of collaborative model.

Based on the research conducted, it may be equally valid to suggest that CSA initiatives are more likely to be integrated into larger farms as part of their strategies for product diversification and market segmentation. While such CSA operations meet the basic criteria outlined in the previous chapter, one might question whether they contradict the original essence of the CSA model—namely, supporting small-scale farmers.

From an economic perspective, the efficiency of CSA operations depends on both their production capacity and the number of consumers involved in the system. A key factor is determining the minimum number of members required to ensure financial stability and enable the farm's further development. To estimate this parameter, a simplified opportunity cost simulation was applied, based on prevailing wage levels in Poland. The analysis took into account both the gross minimum wage and the average gross monthly salary in the enterprise sector, which can serve as benchmarks for assessing the income necessary to sustain CSA operations.

According to the data presented in Table 4.4, the minimum gross annual income per capita in Poland amounts to PLN 50,904, which corresponds to PLN 4,242 per month. Meanwhile, the average gross annual salary in the enterprise sector in 2025 is estimated at PLN 100,896, translating to an average monthly income of PLN 8,404. Comparing these figures helps determine the income levels a CSA farm must generate to remain profitable, as well as the number of consumers it must serve to cover operating costs and labor compensation.

Assuming that the price of one box containing agricultural products is PLN 80 (based on data from wspierajrolnictwo.pl and rwsmarianka.pl) and that the delivery season spans approximately 23 weeks—resulting in the sale of 23 boxes per consumer—the annual revenue generated per

Table 4.3 Characteristics of the Surveyed CSA Farms in Poland

CSA Name	*Is CSA Your Only Source of Income?*	*The Scope of CSA Activities*	*CSA Area (ha)*	*Farm Area (ha)*	*How Many Consumers Are Involved in Your CSA?*	*How Many Consumers Could Potentially Participate in Your CSA at Most?*
CSA Nad Bugiem	No	Vegetables, fruits, eggs, preserves, groats and flours	2.0	4.0	30	60
CSA Zielone Zagrody	No	Vegetables, fruits, eggs, dairy products, meat, preserves	4.0	41.0	60	80
CSA Zielona Rzodkiewka	No	Vegetables, preserves, herbs	0.2	4.8	20	40
Farma Stary Jesion	No	Vegetables, fruits, eggs, preserves, herbs	2.0	36.0	30	60
CSA Marianka	Yes	Vegetables	1.2	4.2	120	120
Green leaf GR Kąkolewice	Yes	Vegetables, eggs, meat, herbs	1.0	13.0	70	80

Source: Author's own elaboration based on survey research.

Table 4.4 Minimum and Average Earnings in CSA Farms

	Gross Monthly Earnings (PLN)	*Number of Months*	*Gross Annual Earnings (PLN)*
Minimum Wage in Poland	4,242	12	50,904
Average Monthly Earnings in the Enterprise Sector (Excluding Profit-Related Bonuses)	8,408	12	100,896

Source: Author's own elaboration based on Statista (2025).

consumer amounts to PLN 1,840. On this basis, it can be estimated that, to achieve revenue equivalent to the minimum gross annual wage, a farmer would need to serve at least 28 consumers. To reach revenue equivalent to the average salary in the enterprise sector, this number would need to rise to 55 consumers.

However, it is important to emphasize that this refers to revenue, not to the farmer's actual income. Revenue does not account for the operational costs of running a farm, such as the purchase of seeds, plant protection products, fertilizers, equipment maintenance, transportation expenses, or labor—whether family or hired. In practice, for a farm to be profitable and financially sustainable, the number of required consumers may be significantly higher than what these basic revenue-based calculations suggest (Prymon, 2019). Moreover, variations in production costs between farms may mean that some require an even larger customer base to achieve satisfactory financial outcomes. Attention should also be drawn to the issue discussed in an earlier chapter of this book, namely the separation between the current operations of a farm and its long-term development perspective. It appears that, as in the case of the aforementioned American farmers, the prices charged by Polish farms likewise fail to reflect the costs of asset depreciation and the need for capital investment. Such an approach will ultimately weaken the economic position of farms and, consequently, does not guarantee the durability of a model that is crucial from the perspective of sustainable development (Raftowicz and Struś, 2026).

The research findings indicate that the six CSA farms analyzed collectively serve 330 consumers. The highest number of customers is served by the farms for which CSA constitutes the main source of income. In particular, two CSAs stand out in terms of customer base: RWS Marianka supplies vegetables to 120 consumers, while RWS Green Leaf GR Kąkolewice serves 70 customers, offering not only vegetables but also eggs, meat, and herbs. Notably, these large customer bases are managed with relatively small cultivation areas—just 1.2 ha in the case of RWS Marianka and 1.0 ha in RWS Green Leaf GR Kąkolewice. This demonstrates that effective CSA operations do not require extensive land, making this food distribution model especially attractive for small-scale farms.

Consequently, from an economic standpoint, CSA appears to be a viable solution for small producers, enabling income stabilization and minimizing market risk through direct cooperation with consumers. However, price determination should be preceded by a thorough cost analysis, including those related to investment activities, allowing at least for the replacement of the fixed assets at the farmer's disposal.

Among the remaining CSA farms, the number of consumers ranges from 20 to 60. Based on earlier simulations, it can be concluded that CSAs with 20–30 members operate at the threshold of profitability. Turning CSA into the sole source of income would require increasing the number of consumers—something that, as shown in Table 4.3, is achievable.

Analysis of the research results shows that limited land area dedicated to CSA does not constitute a direct barrier to development. However, the nature of CSA operations presents significant challenges. CSA farms rely on traditional cultivation methods, which are considerably more labor-intensive than conventional agriculture. At the same time, the limited scale of production means that such farms typically cannot afford to hire additional labor. As a result, most responsibilities fall on the farmers and their family members, which may lead to overwork and constrain development opportunities.

According to the core principles of the CSA model, consumers are expected to actively participate in farm work, supporting farmers in maintaining the operation. However, the study shows that, in practice, such cooperation occurred in only one case—a CSA farm operated year-round by a single individual. In another case, the need for consumer involvement was acknowledged, although no participation in fieldwork was being practiced at the time. In the remaining CSAs, farmers expressed no interest in involving consumers in agricultural tasks, despite simultaneously experiencing labor shortages. This discrepancy reveals a disconnect: on one hand, the lack of labor is a key challenge for CSA operations; on the other, most farmers are reluctant to involve consumers in daily work. This may be due to organizational challenges, concerns about work quality, or the need for increased supervision of individuals unfamiliar with farm practices.

One may also ask whether farmers themselves are genuinely interested in fully implementing the principle of solidarity in practice. It is possible that farmers, in fact, treat CSA primarily as a business model in which they benefit from a form of trade credit—namely, advance payments—while "social responsibility" or the "building of farmer–consumer social ties" remains of secondary importance.

The literature emphasizes that CSA offers numerous benefits to farmers, consumers, and the environment, as discussed in Chapter 2. These findings were compared with the actual motivations of Polish farmers who chose to participate in CSA initiatives, as illustrated in Table 4.5.

Table 4.5 Survey Research Findings

	Market Niche, Higher Product Prices	*Lower Distribution Costs*	*Market Niche, Higher Product Prices Lower Distribution Costs Consumer Interest*	*Prepayments—Limiting Financial Risk*	*Possibility of Obtaining External Support*	*Concern for the Natural Environment*	*Concern for Preserving Traditions and Biodiversity*
CSA Nad Bugiem	Yes	—	—	Yes	—	—	Yes
CSA Zielone Zagrody	—	—	Yes	Yes	—	Yes	—
CSA Zielona Rzodkiewka	—	—	Yes	—	—	—	—
Farma Stary Jesion	—	—	Yes	Yes	—	—	Yes
CSA Marianka	—	Yes	—	Yes	—	Yes	Yes
Green leaf GR Kąkolewice	—	—	—	—	—	Yes	Yes
Total number of affirmative responses	1	1	3	4	0	3	4

Source: Author's own elaboration based on survey research.

Survey results indicate that the main motivation for establishing CSA farms in Poland was the need to minimize farmers' financial risk through a prepayment system, which provides income stability and greater predictability in production. This was highlighted by four respondents. Equally important was the desire to preserve traditional cultivation methods and biodiversity, reflecting a strong commitment among farmers to sustainable agricultural practices. Additionally, three respondents emphasized environmental stewardship as a key reason for engaging in CSA activities.

Based on the responses, it can be concluded that the decision to operate within the CSA model was driven primarily by economic and ecological considerations. Notably, in three cases, the establishment of a CSA farm was also prompted by consumer initiatives—individuals who expressed interest in participating in this model of food distribution. This highlights the growing role of conscious consumers in shaping the direction of modern agriculture and demonstrates that CSA is not solely a farmer-led initiative, but also a product of community collaboration and engagement.

It is worth noting that only one farm reported entering CSA to take advantage of a market niche and the opportunity for higher product prices. Similarly, only one respondent identified reduced distribution costs as a significant factor influencing their decision to participate. These responses suggest that farmers do not view higher prices or intermediary savings as central to the CSA model. Rather, they appear to prioritize income stability and predictability over maximizing unit prices.

Another important observation is that none of the respondents cited access to external support as a motivation for joining CSA. This may reflect the limited involvement of public institutions in promoting and developing this model of agriculture. The absence of significant financial or organizational support discourages farmers from seeing public entities as potential allies—possibly contributing to low institutional interest in CSA and hindering its broader development (Raftowicz and Korabiewski, 2025).

The survey results also reveal several key benefits that farmers associate with CSA participation:

- predictable income,
- guaranteed sales of products,
- time saved on product marketing and more focus on cultivation,
- creation of collaboration networks,
- direct contact with consumers.

From this, it is evident that farmers recognize numerous advantages of participating in short food supply chains. They especially value improved market stability, more accurate income forecasting, and closer relationships with consumers. The CSA model affords them greater independence and enables more sustainable business planning.

Despite favorable conditions for CSA development in Poland—such as the highly fragmented structure of agricultural holdings—the model continues to develop on a limited scale. Its overall impact within the national agricultural sector remains modest, primarily due to a range of barriers hindering its growth and wider adoption.

The survey clearly reveals the existence of diverse barriers to CSA development. These include social and psychological challenges, as well as legal and organizational constraints. One of the most frequently mentioned issues was the difficulty of building cooperation networks. A low level of social capital results in limited trust between farmers and consumers, which is a significant obstacle to the effective implementation of CSA. Trust is fundamental, as the model relies on prepayments from consumers who commit to regularly receiving farm products. It could even be argued that mistrust and reluctance to cooperate represent some of the most critical barriers to the development of this model (Raftowicz and Korabiewski, 2025; Raftowicz and Struś, 2026).

Another major challenge involves regulatory requirements. Farmers noted difficulties in complying with sanitary and veterinary standards, particularly for animal products. These strict regulations may restrict CSA offerings to only fruits and vegetables, significantly narrowing the model's potential reach.

Labor shortages also remain a key concern. Many farmers face difficulties finding adequate labor, which makes it challenging to operate their farms according to CSA principles. In this context, broader consumer participation in fieldwork could be considered, as it may help strengthen relationships between producers and consumers. Paradoxically, environmental care itself can also pose a challenge: the effort to preserve biodiversity and adhere to ecological principles requires substantial labor and time, placing additional burdens on farmers.

Finally, the survey pointed to a lack of appropriate infrastructure in urban areas to facilitate efficient CSA product distribution. Respondents emphasized the need to establish pickup points, which would streamline the food delivery process and increase the overall appeal of the CSA model.

Despite numerous barriers, farmers nonetheless perceive opportunities for the further development of CSA (Community-Supported Agriculture) in Poland. Among the key factors that could contribute to the popularization of this model, they identify the growth of social capital, improvements in dietary habits across society, and increased consumer awareness regarding the origins of food. Education is seen as playing a crucial role in this process and should be directed toward both farmers and consumers. Equally important is support in the area of modern sales tools—farmers emphasize the need for the development of online purchasing platforms and marketing services to facilitate the promotion and sale of products.

The analysis of responses regarding the entities currently supporting CSA—as well as those that farmers believe should become more actively involved in its development—provides valuable insights. The findings indicate that the most significant support presently comes from non-governmental organizations, including those representing producers and those advocating for consumers. However, farmers point to insufficient involvement from both local and national government authorities. They expect greater engagement from local government units, Agricultural Advisory Centers, and District Veterinary Inspectorates in supporting CSA-related initiatives.

In conclusion, despite ongoing challenges, farmers assess the prospects for CSA development in Poland as relatively positive. Critical to the continued growth of this model will be the removal of administrative barriers, the building of social trust, and increased support for farmers through educational, marketing, and infrastructure-related efforts. Collaboration among the public sector, non-governmental organizations, and CSA participants themselves can foster the expansion of the CSA model and ensure its lasting integration into the fabric of Polish agriculture.

4.3 CSA from the Consumers' Perspective

In an effort to analyze Community-Supported Agriculture from the demand side, a survey was conducted among employees of higher education institutions in Poland. As previously mentioned, the selection of respondents was purposeful, based on the assumption that, compared to the general population, they possess greater knowledge of healthy eating and therefore represent a potential group of CSA consumers. The selection of respondents was also consistent with findings from studies conducted in Germany, which—as noted earlier—show that a representative CSA participant is typically a member of the urban middle class (well educated, with relatively high income and strong environmental awareness).

A total of 176 individuals participated in the study, nearly 67% of whom identified as women. The average age of respondents was slightly over 46 years. Approximately 61.4% were academic, research-and-teaching, or teaching staff; 27.8% were administrative staff, and 10.8% were technical employees. The majority of respondents—63.6%—lived in cities with a population of over 500,000. A significant portion—22.2%—lived in rural areas. However, in this case, we are dealing with individuals who, as a result of urban sprawl (suburbanization), have settled in peri-urban areas.

Most responses came from the Dolnośląskie voivodeship, with additional respondents from Kujawsko-Pomorskie, Warmińsko-Mazurskie,

Mazowieckie, Łódzkie, Wielkopolskie, Lubelskie, and Małopolskie. In total, respondents from half of Poland's voivodeships were represented in the study.

For nearly half of the respondents, the net monthly household income per capita fell within the range of PLN 3,000 to PLN 6,000, as illustrated in Figure 4.2.

The research revealed that almost 61% of respondents were familiar with the concept of CSA, indicating a growing awareness of this model of collaboration between consumers and farmers and its importance in the sustainable development of local food production.

Respondents most commonly gained their knowledge about CSA from the internet (websites, news portals, social media—51.1% of responses) and from people with experience in the topic (e.g., family, friends, colleagues—29%). An important source of knowledge diffusion was also scientific publications (books, professional articles—around 13.6%), as well as television and radio (9.1%).

At the same time, 54% of respondents stated that they were unaware of local CSA initiatives, indicating that there is still a significant need for promoting and disseminating this model within local communities.

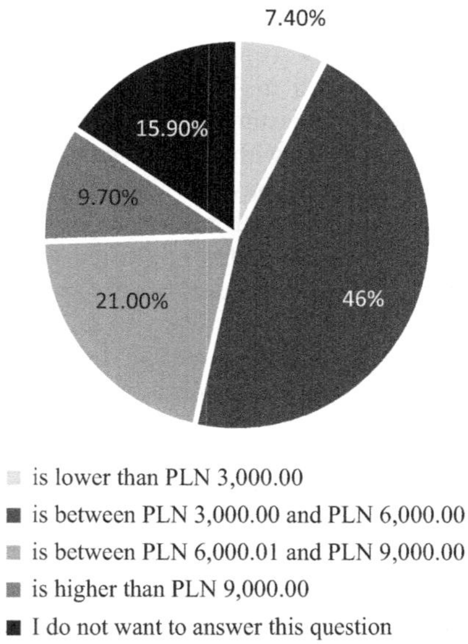

Figure 4.2 Net Monthly Income Per Capita in the Surveyed Household.

Source: Author's own elaboration based on the conducted research.

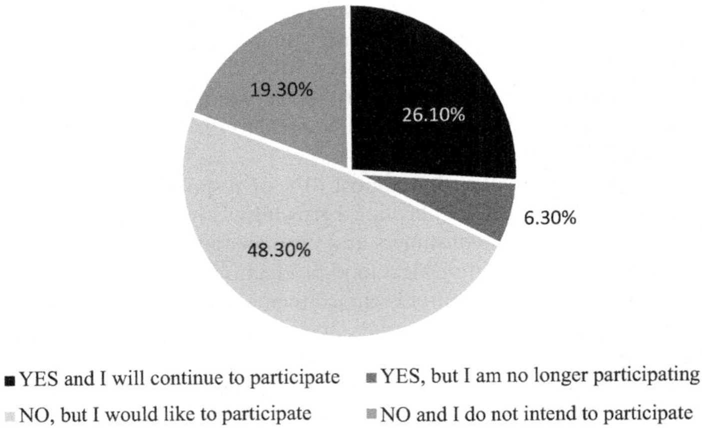

Figure 4.3 Consumer Participation in CSA.

Source: Author's own elaboration based on the conducted research.

It is also noteworthy that respondents had a positive attitude toward participating in CSA initiatives, as illustrated in Figure 4.3.

Over 26% of the respondents participate in CSA and express a desire to continue their participation. At the same time, nearly 49% of respondents are potentially interested in joining CSA, while just under 19% expressed a different view. Interestingly, among the respondents were also individuals who had previously withdrawn from CSA participation (over 6%). The relatively high level of consumer interest in participating in CSA indicates that this model of social innovation has growth potential—provided it can be adapted to the realities of a post-socialist society, where individualism still tends to prevail over solidarity and collective action for the common good.

An interesting conclusion can be drawn from the analysis of the research results regarding consumer associations with the concept of CSA. Most respondents associated CSA with values such as ecological sustainability, healthy and high-quality food, local production, and close cooperation between farmers and consumers. This may indirectly reflect consumer expectations related to potential participation in CSA. Consumers primarily expect farmers to provide organic food produced using traditional methods. At the same time, CSA does not evoke strong associations with solidarity. As a result, it can be expected that consumers may not be particularly interested in sharing risks with farmers through the prepayment system. It should therefore be recognized that, in the perception of Polish consumers, environmental considerations and the local character of production are of primary importance. This perspective should be taken into account when redesigning the CSA model so that it can move beyond the niche in which it currently operates.

The aforementioned willingness of consumers to participate in CSA results from the fact that, for the majority, they consider this model of collaboration beneficial to them. This view was shared by nearly 72% of respondents, as illustrated in Figure 4.4.

It is worth noting that only 0.6% of respondents consider CSA an unfavorable solution for them. Comparing this data with that from Figure 4.1, one could argue that the decision to withdraw from CSA, in the vast majority of cases, did not stem from negative consumer experiences, but rather from other external factors unrelated to the model itself.

Regarding the advantages of CSA, the majority of respondents pointed to direct access to fresh, local products (88.1%), support for local farmers and small farms (77.3%), promotion of a healthy lifestyle and conscious eating habits (52.8%), and the building of social bonds and connections between consumers and farmers (52.3%). The following benefits were also mentioned: supporting sustainable development and organic farming (42.6%), boosting the local economy and creating jobs (35.8%), reducing food transport emissions (29.5%), transparency in the food production process (27.8%), and education on agriculture, ecology, and sustainable development (25.6%). What is worth emphasizing is that the results obtained in this question do not always correspond with the answers provided elsewhere in the survey. For example, consumers identify

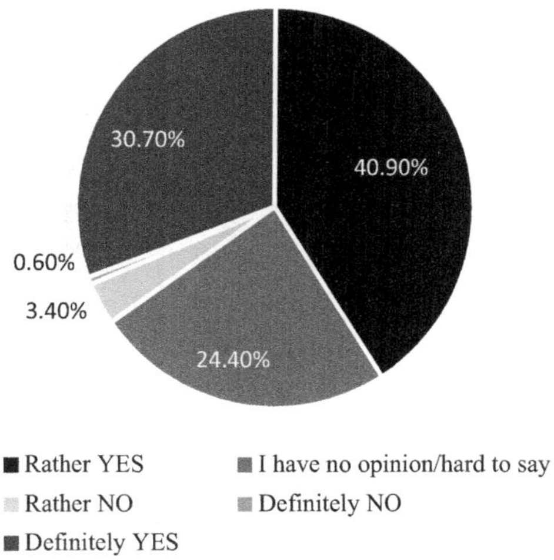

Figure 4.4 Consumers' Evaluation of the CSA Model.

Source: Author's own elaboration based on the conducted research.

supporting local farmers and fostering social ties as advantages of CSA; however, as further questions reveal, they are not particularly willing to undertake concrete actions to support farmers—for instance, through financial contributions or on-farm work. This indicates a clear discrepancy between "theory" and "practice": declarative knowledge does not translate into actual behavior. It appears that this consumer attitude is one of the reasons for the weakness of the classical CSA model and the difficulties associated with its implementation in Poland.

Despite the fact that consumers are aware of the advantages of CSA, they remain unwilling to go beyond the simple model of food sales. This is reflected in consumers' attitudes toward one of the pillars of the classical (Western European) CSA model: consumer participation in farm work. Only 6.3% of respondents expressed a willingness to help farmers on the farm, while a further 21% of respondents considered this possibility (Figure 4.5).

At the same time, 37.5% of respondents do not want to participate in such tasks, or rather, are reluctant to do so. It is worth noting that, as indicated by previous studies, farmers also tend to be skeptical about consumer help in farm work. This leads to the conclusion that this principle does not work well in the Polish context and should be reconsidered.

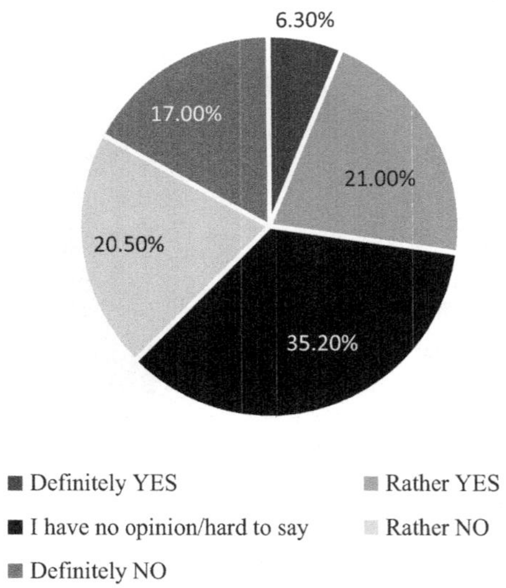

Figure 4.5 Consumers' Willingness to Assist Farmers within the CSA Framework.

Source: Author's own elaboration based on the conducted research.

Respondents also identified several disadvantages of CSA, especially:

- the risk of insufficient production in the case of poor yields (44.3%);
- dependence on unpredictable weather and natural conditions (41.5%);
- lack of flexibility in choosing products within the subscription (35.8%);
- limited variety of available products (31.8%);
- limited availability of products during the season (30.7%);
- issues with timely product collection (30.1%);
- high prices compared to supermarkets (28.4%).

Nevertheless, the vast majority of respondents are interested in participating in CSA in the future, as illustrated in Figure 4.6.

A factor that could encourage respondents to participate in CSA is primarily access to fresh, local, and organic products (64.2%). Other significant factors include support for local farmers and small farms (46.6%) and access to local products with higher nutritional value (44.3%). Price, which could be competitive compared to traditional organic product stores, is also important (45.5%). Other noteworthy factors include the desire to support a healthy lifestyle and diet (36.4%), availability of various subscription options and flexibility in deliveries (24.7%), offering products that are harder to find in supermarkets (22.7%), a low subscription price (20.5%), and shared responsibility for local development (17%).

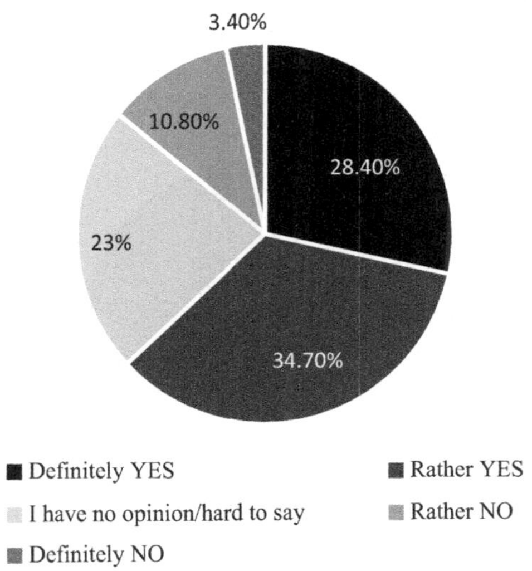

Figure 4.6 Interest in Future Participation in CSA.

Source: Author's own elaboration based on the conducted research.

To summarize this part of the study, it can be stated that consumers are generally interested in participating in CSA, but not necessarily in its traditional form. It appears that consumers expect primarily business-oriented rather than socially oriented relationships with farmers. They want guaranteed access—secured through a subscription—to high-quality organic and traditionally produced food. The declared idea of solidarity and support for farmers is, in practice, reduced to purchasing from local producers. In this context, the prevailing mindset could be described as: "I buy locally, therefore I support local farmers, and thus I am showing solidarity with them."

As previously mentioned, the potential for CSA development stems, among other things, from changes in consumer eating habits. Consumers often perceive a link between the food products they purchase and the occurrence of diet-related diseases. As a result, according to the research, nearly 35% of respondents buy organic products every week, and another 31.3% purchase them once a month (Figure 4.7).

Just under 6% of respondents do not purchase organic products. Organic products are most commonly bought at farmers' markets and bazaars (28.9%) and in supermarkets (organic product section) (25.3%). Less frequently, consumers purchase these products directly from farmers or organic farms (9.6%), in organic product stores (8.4%), or in online stores specializing in organic products (8.4%).

However, when consumers do purchase products directly from farmers, they typically do so 1–2 times a year (about 39%) or once a month (about 20%). Only 11.4% of respondents buy products directly from farmers every week, as illustrated in Figure 4.8.

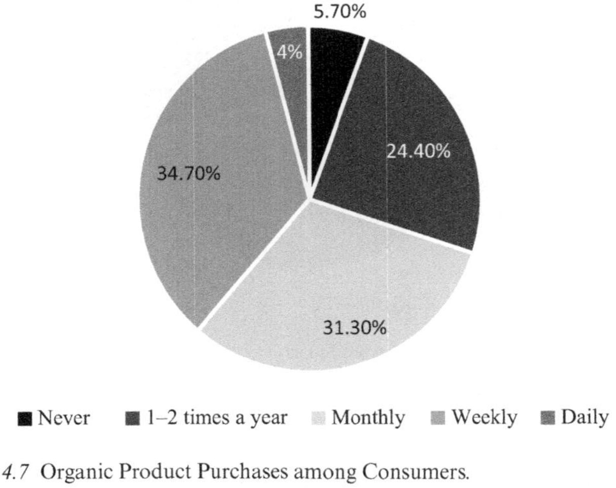

Figure 4.7 Organic Product Purchases among Consumers.

Source: Author's own elaboration based on the conducted research.

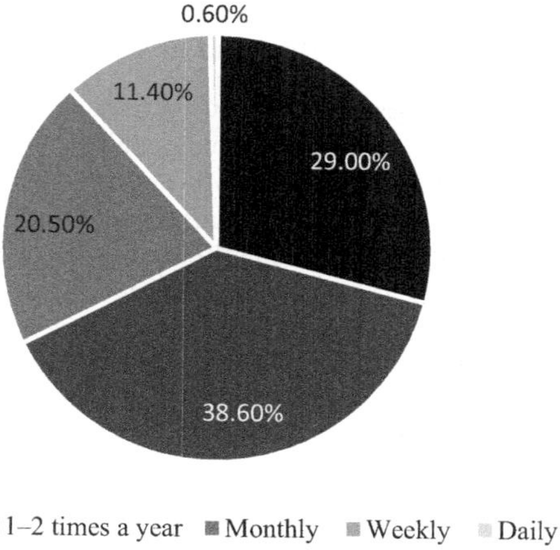

Figure 4.8 Direct Purchases from Farmers among Consumers.

Source: Author's own elaboration based on the conducted research.

In summary, the results of the survey conducted indicate that, in principle, consumers are interested in participating in CSA. They recognize its advantages and the benefits they can derive from it. However, they approach this model rather selectively. Primarily, they are interested in access to organic and traditional food. At the same time, the majority are skeptical about helping farmers with farm work and view the prepayment system more as a barrier than as an advantage of the model. It can thus be concluded that they treat the principle of solidarity more as a declarative concept, and in practice, they are not interested in putting it into practice.

In conclusion, it can be stated that the Western European CSA model, under Polish conditions, will not develop (in quantitative terms) enough to become a real alternative to industrial agriculture. The absence of such development, in turn, will not improve local food security nor make the local food system more resilient to external disruptions. It appears that the fundamental problem among consumers, too, is the low level of social capital—manifested in pronounced individualism, distrust, and reluctance toward institutionalized forms of cooperation, all of which form part of the legacy of the socialist economy. On a positive note, consumers recognize the need to create alternatives to industrial agriculture. The vision of sustainable agriculture, particularly its environmental dimension, is not unfamiliar to them. Consumers are willing to support farmers, but primarily in an economic sense—by generating demand for their products and purchasing them directly from producers.

4.4 The Role of Agricultural Support Institutions in Poland in the Development of CSA

The final part of the research focused on the role of Agricultural Support Institutions in promoting short supply chains. As part of this analysis, studies were conducted within the Ministry of Agriculture, specifically in the Department of Agricultural Markets and Rural Energy Transition, which is responsible for organizing the country's agri-food supply systems.

The findings indicate that the Ministry lacks a unified and effective strategy to significantly encourage farmers to develop short supply chains. Despite the implementation of legal regulations in 2017 aimed at facilitating such initiatives, and the introduction of financial support instruments in 2019, the effects of these efforts have remained limited. The Ministry highlights difficulties in cooperating with farmers, which stem in part from mental barriers and a low willingness to engage in collective initiatives within organized distribution systems. Currently, the Ministry is undertaking several actions to support short supply chains, including the launch of the e-bazarek online platform, funding rural women's associations, and organizing monthly markets on its premises to enable direct sales by local producers. However, the scale of these initiatives remains insufficient, and the lack of a comprehensive approach means that the short supply chain model is developing slowly.

The research also found that the Ministry is not actively promoting the development of Community Supported Agriculture. Rather than implementing a long-term strategy inspired by international best practices, the initiatives undertaken are ad hoc and not always well thought through. One example was a 2023 government proposal to introduce special shelves in hypermarkets for food from local suppliers, aimed at increasing the availability of regional products. Although the initiative initially drew significant public interest, it quickly faced criticism—from retail chains, which pointed to logistical challenges, and from farmers, who feared price pressure imposed by large supermarkets. Ultimately, the project was abandoned, reinforcing the impression of an incoherent policy in supporting local food systems.

Similar research was conducted at the National Support Centre for Agriculture (Krajowy Ośrodek Wsparcia Rolnictwa—KOWR), specifically within the Department of Market Development. One of the key statutory responsibilities of this institution is to carry out and support promotional, educational, informational, and cultural activities focused on agricultural and food products, their production methods, and quality systems. Additionally, KOWR is responsible for managing promotional funds for agri-food products, aimed at supporting producers and increasing consumer awareness of their goods.

The research findings indicate that KOWR primarily focuses on organizing various promotional events, such as fairs, exhibitions, demonstrations, and competitions. However, most of these initiatives are targeted mainly at consumers, rather than directly at farmers and producers. A central aspect of the Centre's activities is education and raising public awareness of the quality and origin of food. As a result, KOWR places strong emphasis on the development of culinary tourism, the organization of festivals dedicated to regional products, and the implementation of informational campaigns promoting traditional and local foods.

According to KOWR representatives, the development of CSA in Poland is closely linked to consumer purchasing decisions, and consumers should be appropriately encouraged to choose local products. As part of this strategy, the institution is developing a calendar of culinary events aimed at popularizing regional specialties and increasing public interest in domestically produced food. Additionally, KOWR operates a specialized website that serves as an informational platform, providing consumers with easier access to data about local producers and their offerings.

Despite its ambitious goals and broad range of activities, the analysis of KOWR's operations shows that the institution focuses more on end consumers than on supporting producers in developing local distribution systems. Moreover, the research revealed that KOWR does not actively promote the idea of CSA, which poses a significant barrier to implementing internationally proven cooperation models. The lack of awareness and promotion of this system limits opportunities to strengthen the position of farmers and producers, who could benefit from more efficient and stable forms of direct sales.

An important part of the research also involved interviews conducted at the Agricultural Advisory Centre in Brwinów and at regional Agricultural Advisory Centers. The Agricultural Advisory Center (CDR) in Brwinów acts as a national unit supporting the agricultural advisory system. It coordinates, trains, and provides methodological support to the regional Agricultural Advisory Centers (ODR), although it does not supervise them. It is involved in implementing innovations, international cooperation, and delivering both EU and national programs. CDR focuses on agricultural development, farmer education, and the promotion of modern technologies and sustainable development within the agricultural sector. The analysis of the collected responses revealed significant regional differences, reflecting a varied approach to the topic and differing conditions for advisory services across the country, as illustrated in Table 4.6.

The analysis of the research revealed significant differences in the level of CSA system development across individual voivodeships, as perceived by the Agricultural Advisory Centers. The lowest rating for the system's

Table 4.6 Assessment of the Functioning of CSA in Selected Voivodeships by the Regional Agricultural Advisory Centers and the Agricultural Advisory Center (CDR) in Brwinów

	Assessment of the Functioning of the CSA System	Potential for Further Development of CSA	The Role of Institutions in Promoting CSA	Level of Formal Cooperation among Farmers	Level of Informal Cooperation among Farmers
mazowieckie	3	5	5	3	4
wielkopolskie	3	4	5	4	5
łódzkie	4	4	5	4	3
zachodniopomorskie	1	1	5	3	3
CDR	4	5	5	3	4

(1 means very poor and 5 means very good).
Source: Author's own elaboration based on the conducted research.

functioning was given to the Zachodniopomorskie Voivodeship, while the best results were observed in the Łódzkie Voivodeship. The highest development potential for the CSA system was identified by representatives of the Mazowieckie Voivodeship and experts from the Agricultural Advisory Centre (CDR) in Brwinów, who awarded it the highest possible score (5). This indicates favorable conditions for further CSA expansion in this region, likely due to the presence of Warsaw—the largest city in Poland by population (1.86 million residents). The capital stands out with the highest average wages in the country and a rapidly growing urban middle class, which often demonstrates higher environmental and social awareness. Therefore, it can be assumed that the high rating of CSA development potential in the Mazowieckie Voivodeship is primarily linked to strong demand and its anticipated growth.

The situation in other voivodeships varies. Particularly concerning is the low level of CSA development in the Zachodniopomorskie Voivodeship (score 1), indicating a need to intensify support efforts. Despite differences in performance levels, all analyzed voivodeships received the highest score (5) for institutional engagement in promoting the CSA system. This suggests that the entities responsible for CSA support rate their own efforts highly; however, this does not always translate into actual cooperation with farmers.

The assessment of formal cooperation among farmers in most voivodeships was at a moderate level (3–4), while informal cooperation proved to be more developed—especially in the Wielkopolskie Voivodeship, which received the maximum score (5). Based on these results, it can be concluded that CSA development largely relies on interpersonal relationships, while institutional support remains insufficient. Given the relatively small number of CSA initiatives currently operating in Poland, the lack of effective systemic solutions poses a significant barrier to the broader adoption of this model.

A standout institution in terms of CSA support is the Agricultural Advisory Center (CDR) in Brwinów, which rates its activities highly in both the implementation and promotion of CSA-related initiatives.

At the local level, the research was conducted in six municipalities where CSA initiatives were active:

- Mazowieckie Voivodeship: Jabłonna Lacka (rural municipality)
- Mazowieckie Voivodeship: Słubice (rural municipality)
- Wielkopolskie Voivodeship: Łobżenica (urban-rural municipality)
- Wielkopolskie Voivodeship: Budzyń (urban-rural municipality)
- Łódzkie Voivodeship: Aleksandrów (rural municipality)
- Zachodniopomorskie Voivodeship: Będzino (rural municipality)

Detailed research results are presented in Table 4.7.

Table 4.7 Results of CSA Research Conducted in Selected Municipalities

Municipality	*Assessment of the Functioning of the RWS/LRR System*	*Potential for Further Development of the CSA/Local Agricultural Market in the Commune*	*The Role of the Commune in Promoting the CSA/Local Agricultural Market*	*Which Institution Should Support the Development of RWS/Local Agricultural Market?*
Jabłonna Lacka	Lack of knowledge about the activities of the RWS in the commune	Minimal	Minimal. It is involved in the harvest festival, there are rural women's associations	Central authority
Słubice	Lack of knowledge about the activities of the RWS in the commune	Minimal—lack of interest of residents in local food	Minimal. It is involved in the harvest festival, there are rural women's associations	Central authority
Łobżenica	Lack of knowledge about the activities of the RWS in the commune	Dormant—due to the mentality of residents and reluctance to collective actions	Minimal. Preferential prices at the market for local producers, no promotion of local food	NGO
Budzyń	Awareness of the idea, but lack of RWS/LRR initiatives	Minimal—wealthy commune, focused on industry	Minimal. Limited local trade, administrative problems	Mayor, if there is interest
Aleksandrów	Awareness of the idea, but limited actions	Medium—lack of financial resources	Minimal. It supports rural women's associations, but there are no markets	Municipality, if it has financial resources
Będzino	Functioning limited to individual events	Large—presence of informal local producers	Facilities at the market for local producers	ODR

Source: Author's own elaboration based on the conducted research.

The research results indicate that the level of knowledge and degree of engagement of municipal authorities in promoting and supporting CSA-related initiatives are generally low. In most of the municipalities analyzed, there is a lack of awareness regarding the operation of CSA systems in their areas and of local agricultural markets. Only a few local governments undertake any actions to support their development. Most municipal leaders do not recognize the potential of this model of cooperation between farmers and consumers, with the notable exception of the Będzino municipality, which is one of the few showing greater interest in this form of direct sales.

Although some municipalities possess favorable conditions for CSA development—such as the presence of local agricultural producers or growing consumer interest in farm-to-table food—this model still faces numerous challenges. The main barriers include: a lack of appropriate infrastructure for efficient distribution, limited financial resources to support such initiatives, and the mindset of residents, who are often skeptical of alternative forms of food purchasing.

The activities of local authorities tend to focus mainly on organizing cultural events, such as harvest festivals, and supporting the work of rural women's associations. While these initiatives play an important role in fostering community integration, they do not significantly contribute to the promotion or implementation of short food supply chains. An additional challenge lies in concerns about bureaucracy and complex legal requirements—particularly related to sanitary and veterinary oversight—which effectively discourage producers from formally participating in CSA systems.

The research also found that municipal authorities show little initiative in developing institutional solutions to support CSA. Support for the development of these systems is most often seen as the responsibility of the central government, NGOs, and agricultural advisory centers. The lack of clear commitment from local governments to build stable support mechanisms for CSA represents a significant obstacle to the growth of this form of cooperation, and ultimately limits the effective implementation of short food supply chains at the local level.

To summarize the research findings: under current conditions, there is no coherent or comprehensive strategy for CSA system development. Actions in this area are fragmented and stem from grassroots initiatives by both farmers and consumers seeking to establish direct cooperation. CSA initiatives in Poland do not receive sufficient support from public institutions and lack consistent support mechanisms at the national level. As a result, CSA development in Poland relies largely on informal relationships and personal contacts between participants, which limits both its scale and long-term sustainability.

The study also highlighted several key barriers to CSA development, including insufficient education about this model of cooperation, inadequate infrastructure to support effective system functioning, and limited access to financial resources that could help farmers and consumers build lasting CSA structures. However, examples from municipalities such as Będzino and Łobżenica demonstrate that local initiatives—such as organizing markets or building informal networks—can provide a solid foundation for further developing short food supply chains.

Unfortunately, the lack of a strategic approach and systemic support from public institutions significantly limits the potential of CSA in Poland. For this model to develop sustainably and over the long term, it is essential to implement well-designed measures including education, infrastructure development, and financial and legal support mechanisms that will enable the stable operation of this form of cooperation on a broader scale.

5 Recommendations and Future Directions for CSA Development

5.1 Opportunities to Increase Community Engagement in CSA

As the research has shown, both farmers and representatives of public authorities expect greater involvement from non-governmental organizations (NGOs) in the development of CSA systems. In theory, these organizations could initiate projects aimed at establishing CSA initiatives, as well as support their ongoing operations. They could also organize and conduct training for farmers interested in participating in CSA. Equally important could be efforts to promote the idea of CSA among consumers.

The role of NGOs in the Polish context is particularly important, as—already noted—Poland faces a low level of social capital, which hinders the formation of relationships between farmers and consumers. There is also a lack of local leaders with the knowledge and determination needed to connect producers and consumers. Based on the existing research, it can be concluded that NGOs may serve as an effective tool for the efficient implementation of selected public tasks. They are a key component of local socio-economic development and, through collective action, help integrate local communities (Kołodziejczyk et al., 2015). Participation in civil society organizations is voluntary and characterized by collective action aimed at promoting mutual benefits (Olson, 1965).

In the "old" European Union countries, NGOs had the time to evolve and develop properly. According to F.A. Hayek (2007, p. 65), these organizations developed spontaneously and organically, as part of an evolutionary process. Consequently, they are now better equipped to carry out public tasks, including those that support CSA systems. At the same time, these organizations continue to build trust among local residents, thereby continually strengthening social capital (van Oers et al., 2018).

In Poland—as in other Central and Eastern European countries—the situation is different. Due to historical circumstances, the development of

DOI: 10.4324/9781003745198-6

NGOs was more revolutionary than evolutionary, and many organizations continue to struggle with limited financial resources as well as persistent mental barriers rooted in the communist era.

In the countries of Central and Eastern Europe, including Poland, the development of initiatives based on cooperation and trust faces specific constraints stemming from the legacy of state socialism. As Howard (2003) notes, post-communist societies are characterized by weak civic structures, low levels of participation in social organizations and a deficit of institutional trust. Sztompka (1999) emphasizes that the lack of trust and the tendency toward the atomization of social life constitute significant barriers to the development of cooperation-based relations. Consequently, civic engagement in initiatives such as CSA requires not only organizational support but also a long-term and systematic effort to build a culture of trust and shared responsibility.

In recent years, however, positive changes have been observed: bottom-up initiatives and non-governmental organizations have begun to emerge, which—through education, local community engagement and joint activities centered around food—contribute to rebuilding trust and social capital in rural areas (Zajda and Mazurek, 2022). Collaboration between local NGOs and farmers involved in CSA can serve as an important mechanism for involving communities in decision-making processes and enhancing the long-term sustainability of such initiatives.

As a result, in contrast to mature democracies, where NGOs are most often created bottom-up, spontaneously and autonomously, in countries where democracy was re-established after 1989—including Poland—NGOs are often established in a top-down manner and tied to specific projects. This means that the goal gives rise to the organization, rather than the organization striving toward a goal. Consequently, NGOs frequently depend on guaranteed financial resources (e.g., EU funds), and their activity tends to end once the funding or project concludes (Raftowicz-Filipkiewicz et al., 2016).

This situation may at least partially explain why NGOs currently do not engage to a satisfactory degree in the functioning of CSA. The lack of dedicated CSA funding means there is no financial incentive for NGOs to become involved in such projects. Although this behavior undermines the very essence of grassroots social initiatives, in post-communist conditions, it remains a reality that cannot be ignored. Moreover, one might argue that although NGOs in relatively young democracies resemble market actors, the creation of purpose-specific entities (established to carry out a particular task) is natural in a market economy and should not be criticized (Raftowicz-Filipkiewicz et al., 2016).

Despite these concerns regarding the functioning of NGOs in Poland and their role in supporting CSA, some positive examples can also be cited. In Poland, the following NGOs are involved in promoting CSA-related ideas:

- AgriNatura Foundation in Świerże-Panki village
- Institute for Global Responsibility in Warsaw
- Rural Development Foundation in Warsaw
- EcoDevelopment Foundation in Wrocław
- Open Plan Foundation in Kraków

It should be noted that this list is far from complete and includes only those organizations that are more actively engaged in CSA-related work. It's also worth noting that the scope of these organizations' activities is broader and primarily focused on environmental issues in a general sense. Although CSA is an important topic, it typically represents only a fragment of their overall mission—with the exception of the AgriNatura Foundation, for which CSA is a core focus.

An analysis of these organizations' activities indicates that their efforts are mainly directed toward raising awareness about CSA. They also often provide training for individuals (both farmers and consumers) interested in participating in CSA. However, it should be emphasized that such training is typically one-off in nature. There is a lack of regularly scheduled training programs and consistent legal, economic, and financial advisory services. These organizations also rarely conduct farmer training in rural areas—that is, on-site and in local contexts. This is largely due to the financial constraints of NGOs and their dependence on funds obtained through specific, often short-term EU or national projects.

In rural areas, which often struggle with low levels of social and human capital, Local Action Groups (LAGs) should also play an important role. LAGs are a form of territorial partnership established as grassroots initiatives, bringing together representatives of local public, private, and non-governmental sectors, along with citizens acting in the interest of their communities. In Poland, there are 324 such groups (according to KSOW). These entities are meant to integrate local environments and, in doing so, build social capital that could later support the development of CSA. However, at present, LAGs play a limited role in promoting the CSA concept. This may be due to staff shortages (LAGs often employ only a few full-time staff) and the fact that they must obtain funding for promotional activities through grant competitions or contributions from local producers and food processors. Nevertheless, LAGs hold untapped potential that could be used to support the creation and operation of CSA initiatives.

In conclusion, although the current role of NGOs in promoting and implementing the CSA model in Poland is limited, they could become key leaders and initiators of such initiatives in the future. However, for this to happen, NGOs must first develop an operational model that ensures organizational stability, including financial sustainability.

5.2 Policies Supporting the Development of CSA

As the analysis has shown, CSA still represents a relatively small part of the agricultural sector in Central and Eastern European countries. Its development depends largely on favorable institutional conditions and support from public policies. Policies supporting CSA include actions at the EU, national, and local levels. The importance of shortening food supply chains in European Union policy is evidenced by its inclusion in the European Green Deal, which sets the direction for development up to 2050. However, it should be clearly emphasized that, in the Polish context, such measures are of a subsidiary nature. CSA initiatives emerge bottom-up, and the policies supporting them are not of decisive importance either for their formation or for their day-to-day functioning.

Recognizing the need to act on shortening food supply chains, the European Commission adopted two key documents in May 2020: the EU Biodiversity Strategy for 2030: Bringing Nature Back into Our Lives (COM, 2020a, 380) and the Farm to Fork Strategy for a fair, healthy, and environmentally friendly food system (COM, 2020b, 381). The new Biodiversity Strategy aims, by 2030, to address biodiversity loss, unsustainable land and sea use, overexploitation of natural resources, environmental pollution, and invasive alien species. The European Commission estimates that the actions undertaken—backed by a guaranteed €20 billion budget—will deliver economic benefits to local communities, contributing to the creation of sustainable jobs and economic growth.

Meanwhile, the Farm to Fork Strategy envisions a transition to a sustainable food system based on shortening supply chains in EU member states, aiming to ensure food security and access to healthy nutrition. The strategy also seeks to reduce the environmental and climate footprint of EU food systems and to provide livelihoods for economic actors.

According to the European Commission, these two strategies are complementary, linking nature and business—i.e., farmers, companies, and consumers—in a joint effort toward sustainable competitiveness in the future. They also align with the European Green Deal, adopted in December 2019, which sets out an ambitious action plan for more efficient resource use through a transition to a climate-neutral economy, where economic growth is not dependent on resource consumption. The Green Deal also aims to halt biodiversity loss, reduce pollution levels, and transform food systems into models of sustainable competitiveness, public health protection, and livelihood security for all actors in the food value chain (COM, 2019).

Although the European Green Deal is not directly focused on CSA, its core priorities—such as environmental protection, shortening food supply chains, support for small farms, and ecosystem regeneration—are closely aligned with the principles of this model. As a result, CSA initiatives can benefit from EU funding, legislative support, and growing consumer awareness.

At the national level in Poland, farmers involved in CSA can access various forms of financial support, although there is no dedicated grant program exclusively for this model. Potential funding sources include:

- Rural Development Program (PROW): includes support for small farms and initiatives promoting short supply chains and direct sales. CSA farmers can apply for funds to expand their operations, for example, under "Start-up aid for non-agricultural activities" or "Support for investment in the processing of agricultural products."
- Local Action Groups (LAGs): under the LEADER initiative, farmers may apply for grants to promote local agriculture, which can support the CSA model.
- Subsidies for organic farming: CSA farmers operating organic farms may receive payments under EU programs, such as subsidies for organic farming through the Common Agricultural Policy (CAP).

Table 5.1 presents the entities and organizations with the greatest influence on the efficiency of the food supply chain shortening process, which can directly or indirectly contribute to the success of CSA initiatives.

From the perspective of post-communist countries, a key challenge for the development of CSA is the weak coordination between public institutions and civil society organizations. As noted by Zajda and Mazurek (2022), the absence of established cooperation mechanisms between public administration and the civic sector limits the capacity to implement social innovations in rural areas. In contexts where state institutions operated for decades in isolation from local communities, rebuilding trust and partnership requires time as well as appropriate support instruments. At the same time, research by Visser et al. (2019) shows that in many Central and Eastern European countries, small farmers, despite limited institutional support, are developing alternative forms of cooperation—often informal and based on reciprocity and the exchange of goods and services. These practices may serve as a foundation for designing public policies that more accurately reflect the actual conditions under which family farms operate in the post-communist region.

In situations where CSA initiatives are supported by the public sector, their economic efficiency tends to improve even further, as government, quasi-government, or local government institutions have greater market influence than individual farms. However, this requires coordinated

Table 5.1 Institutions Involved in CSA

International Institutions	• Non-governmental organizations (FAO, WHO), • European Commission,
National-Level Institutions	• Ministry of Agriculture, • Legislative and regulatory bodies, • Agency for Restructuring and Modernization of Agriculture, • National Support Center for Agriculture, • National Rural Network, • Educational and research institutions (Institute of Agricultural and Food Economics, State Research Institute),
Regional-Level Institutions	• Marshal's Offices—Rural Departments, • Agricultural Chambers, • Farmers' associations, • Regional associations, • Support services (control bodies, agricultural advisors, educators),
Local-Level Institutions	• Public sector: local authorities, • Private sector: consumer groups, farmers, enterprises, • Local associations, cooperatives,

Source: Author's own elaboration.

efforts among the various agencies responsible for managing such models, in order to eliminate conflicting actions and messages. Transitioning to CSA, in addition to requiring organizational transformation at the farm level, also involves consistent marketing efforts and substantial initial investments.

Therefore, building CSA systems—due to their multidimensional nature—is a long-term and complex process that demands the implementation of coherent, professionally planned, and sustained communication strategies. These strategies should aim to achieve steady market share growth, enhance the appeal of products and services, and strengthen the image of farms and rural areas. Such a strategy should be cross-political, visionary, long-term, and multifaceted, taking into account the specific production and social characteristics of each country or region. In practice, this implies stronger involvement of authorities at all levels of administration in the development of CSA.

In Poland, after initial setbacks in promoting CSA models, attention was drawn to the need for a greater role of local governments in projects initiated by the Ministry of Agriculture and Rural Development. The first step toward effective implementation of the CSA model is overcoming organizational barriers. To achieve this, the municipality must take on the role of an incubator for initiatives aimed at organizing, on one hand, an operational group of producers, and on the other, consumers willing to engage with the proposed solutions, as illustrated in Figure 5.1.

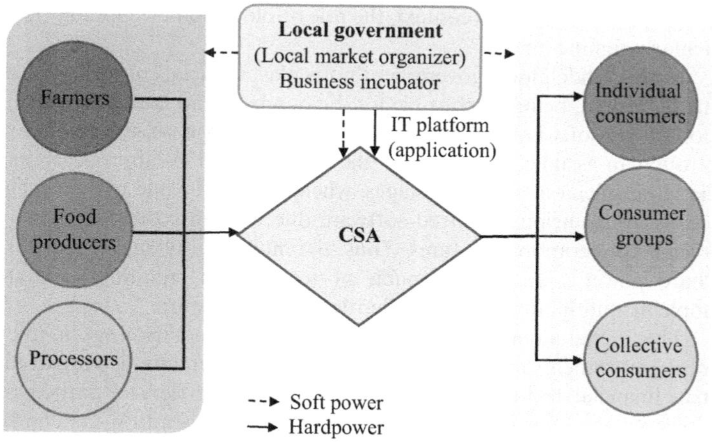

Figure 5.1 CSA Model Involving Local Government.

Source: Author's own elaboration based on: (Kachniarz and Raftowicz, 2020).

In the presented model, the municipality serves as the Local Organizer of CSA (Community-Supported Agriculture), focusing primarily on stimulating supply and demand through supportive initiatives such as promotion, consultancy, integration, and coordination. However, it does not engage directly in economic activities—this role is described in the model as "soft power" (illustrated with a dashed line). The outcomes of these actions can benefit both individual consumers and collective catering institutions managed by local government units, such as canteens in kindergartens, schools, hospitals, or nurseries.

Key economic responsibilities—such as generating revenue, incurring costs, and managing tax and financial settlements—remain solely with the sellers: farmers, food producers, or processors. The model is based on maintaining direct relationships between the supply and demand sides, thereby minimizing the role of intermediaries. To better coordinate activities, achieve synergies, and create attractive purchasing options for customers, producers can form Operational Groups in the form of cooperatives, associations, foundations, or commercial companies. In this structure, the sales function can be separated from the individual farmer or producer and assigned to a dedicated Operational Group.

A vital component of this model is the use of modern technologies, particularly IT systems that optimize and expand collaborative networks. The implementation of digital tools goes beyond order processing and transaction settlements—as in standard e-commerce—and also improves the logistics of connecting producers and consumers across different geographic areas. In this context, the role of blockchain technology is particularly significant.

In this model, local government units play a crucial role in organizing the IT system, being responsible for the purchase and initial implementation of the software. After the initial operational phase, the Operational Group can assume ownership of the technology. This approach is especially beneficial at the early stages, when the supply side may face challenges in acquiring advanced software due to high costs and a lack of formal cooperation structures. Thus, the municipality acts as a form of "hard power"—an entity capable of securing EU funding for system implementation, including both hardware and software.

This model aligns with the broader goal of strengthening the role of local governments in organizing short food supply chains, helping to eliminate financial, technological, and organizational barriers for participants. However, the effectiveness of this solution will depend on the commitment of local government representatives, as the support and development of short food supply chains are not mandatory municipal tasks and are carried out on a voluntary basis. Moreover, municipalities often lack the necessary expertise in working with enterprises, which may pose challenges in implementing such initiatives.

5.3 Forecasts for the Future of CSA in the Context of Socio-Economic Changes

As previous research has shown, CSA in Central and Eastern Europe is developing at a slower and less dynamic pace than in Western Europe, despite responding to socio-economic changes, growing environmental awareness, and the need to build local food systems. In the future, however, greater interest in local and organic food can be expected, driven by increasing consumer awareness of product quality, origin, and environmental impact. Food crises, along with global political and economic instability, are prompting people to seek more stable, local sources of food—making the CSA model increasingly attractive. It can be assumed that CSA will become a viable alternative to mass food production and supermarkets. This is also reflected in developments across EU member states. For example, the German Zukunftskommission Landwirtschaft (2021)—the Commission for the Future of Agriculture—recommended in its 2021 final report that the CSA model be promoted as a means of diversifying German agriculture and strengthening rural areas. The report highlights the innovative organization of food production and consumption within CSA, as well as its social, cooperative, and educational dimensions.

Changes in the economy, including inflation and rising agricultural production costs, are pushing farmers to adopt more sustainable and cost-effective cultivation methods. Rising prices of fertilizers, fuel, and energy have increased the relevance of local supply networks, which help eliminate intermediaries and stabilize prices for both farmers and consumers. Growing support for small-scale farms under EU programs such as the European Green Deal may further foster CSA development, particularly during times of economic uncertainty when direct control over food sources becomes especially important.

Technological advancement and digitalization are also playing an increasingly important role in CSA operations. Applications and platforms that support order management, payments, and logistics are becoming more prevalent, while social media and websites serve as powerful tools for promotion and consumer engagement. Precision agriculture and agroecology allow for more efficient crop management while minimizing environmental impact. Thanks to these technologies, the CSA model may become more accessible even to individuals who have not previously been involved in such collaborative systems.

Political and legislative support is equally essential. The European Union is placing increasing emphasis on small and organic farms, as reflected in the Common Agricultural Policy and local funding programs supporting short supply chains and agricultural initiatives. Simplified procedures for direct sales are making CSA a more viable business model, potentially leading to its integration into national and EU-level agricultural development strategies.

Social changes and the growing need for cooperation may further support the spread of the CSA model. A new generation of consumers and farmers is placing higher value on ethical food production and sustainable development. CSA—based on long-term relationships between producers and consumers—addresses the social need to strengthen local communities. The rising interest in plant-based diets and organic products aligns well with the CSA approach, which can expand not only in rural areas but also in urban settings.

Nonetheless, depopulation and an aging population in Central and Eastern Europe—particularly in Poland—may pose challenges to the future of CSA. A declining number of farmers, the closure of small farms, and rural-to-urban migration may hinder the model's expansion. Without effective support programs for young farmers, the number of CSA operations may decline. In response to these demographic shifts, urban CSA initiatives such as community gardens, urban permaculture, vertical farms, and food cooperatives may grow in importance. The model may increasingly evolve into forms adapted to urban conditions, linking farmers and consumers through local supply chains and partnerships with restaurants and neighborhood stores.

Given the decreasing number of people willing to work in agriculture, automation, and modernization of farming practices may provide viable solutions. Smart irrigation systems, agri-robots, and digital platforms that simplify production management can help maintain the profitability of CSA operations. However, this requires significant investment and access to modern knowledge and technologies. Simultaneously, new support programs targeting young farmers are necessary to encourage the establishment of CSA farms, through attractive grants and legal incentives.

Lifestyle and mobility trends will also impact the future of CSA. As more people lead fast-paced lives and are less willing to commit to long-term obligations, more flexible cooperation models may be required. Traditional membership systems may be replaced by flexible subscriptions and on-demand delivery options. Virtual platforms enabling efficient coordination of orders and deliveries may become key tools in scaling up CSA networks.

Climate change will also play a crucial role in shaping the future of CSA. Extreme weather events such as droughts, floods, and heatwaves can disrupt supply chains and force farmers to adopt new adaptation strategies. These may include introducing more resilient crop varieties, regenerative farming practices, and improved irrigation systems—all of which will be essential for resilience.

Another factor influencing CSA development is globalization and competition from imported food. Inexpensive products available in supermarkets can undermine local farms, making it essential to educate consumers about the benefits of choosing locally sourced food. Government and EU support in the form of tax relief and subsidies for small producers will be key in maintaining CSA's competitiveness.

Finally, the future of CSA will depend significantly on consumer education and awareness. Information campaigns, school-based education programs, and workshops on organic farming can stimulate greater interest in CSA. In the long term, education may become one of the most critical drivers for CSA development—fostering public engagement with local food systems and helping build a more sustainable future.

5.4 The Recommended CSA Model

Based on research conducted among both farmers and consumers, it can be concluded that institutional conditions in Poland constrain the implementation of the classic CSA model observed in Western European countries. This model is founded on four pillars: Partnership, Solidarity, Proximity, and Diversity, as illustrated in Figure 5.2.

It appears that in Poland—and more broadly in Central and Eastern Europe—the CSA model requires redefinition to better align with local conditions shaped by distinct development paths, as illustrated in Figure 5.3

CSA MODEL		
⇧	⇧	⇧
1. PARTNERSHIP	**2. SOLIDARITY**	**3. PROXIMITY**
• **Agreement:** Concluded between consumers and farmers. • **Costs:** Transparently estimated production costs. • **Price:** Determined based on production costs. • **Compensation:** Reflects the labor of the farmers.	• **Shared risk** • **Crop failure and surplus:** Support during difficult periods and sharing ofsurplus in times of abundance. • **Upfront payment:** A key expression of solidarity. • **Consumer participation in farm work**	• **Geographical closeness** enabling direct, personal contact. • **Understanding the realities of food production** • **Supporting the local economy** through conscious purchasing decisions
⇧	⇧	⇧
4. DIVERSITY		
• **Sustainable cultivation model:** Based on crop rotation and species diversity • **Seasonality of products** • **Balanced consumer diet:** Ensured through diverse crop production • **Food security (food quality)**		

Figure 5.2 The Classic CSA Model.

Source: Author's own elaboration.

To begin with, the four core principles of CSA should be critically assessed:

- The Principle of Partnership: This principle appears to be accepted by both farmers and consumers. Informed consumers are willing to pay higher prices for traditional and organic agricultural products to compensate farmers for their greater labor inputs. It is worth noting that, due to the shortened supply chain, these prices may still be lower than those in stores selling organic or traditional food. Moreover, entering into a formal agreement serves the interests of both parties. Such agreements help stabilize the market by ensuring farmers a secure

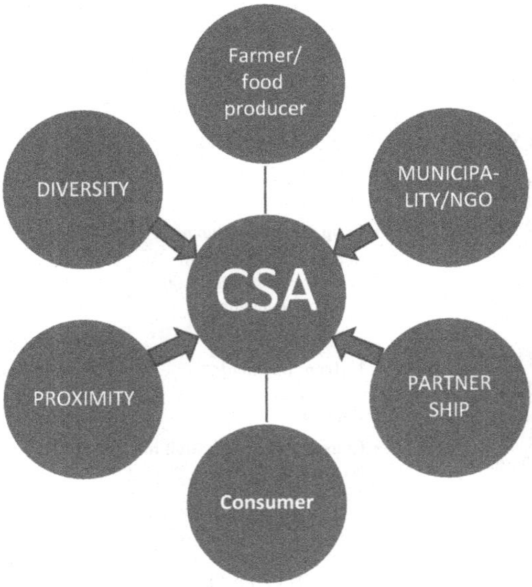

Figure 5.3 The New CSA Model in Poland.

Source: Author's own elaboration.

outlet for their produce and providing consumers with access to trusted sources of food. However, experiences from other countries, such as the United States, show that this principle does not resolve all the challenges faced by farmers. It should be noted that farmers and consumers do not enter into long-term contracts. The price paid by consumers is relatively favorable for farmers and ensures their day-to-day operation, yet it does not allow for investment in the farm. Moreover, in Poland, CSA lacks a regulated legal form, which further hinders long-term cooperation. As a result, risk is indeed shared between producers and consumers, but the time horizon of this risk-sharing is short and typically limited to a single year.

- The Principle of Solidarity: While this principle lies at the core of the model in Western European societies, its implementation in Poland and other Central and Eastern European countries faces challenges. It may be argued that this principle is not fully understood by consumers. Many do not see the need to explicitly support farmers and are therefore reluctant to share the risks associated with agricultural production. This is reflected in their skepticism toward making advance payments for products that may ultimately be delivered in smaller quantities than initially expected. Furthermore, consumers are generally not interested in supporting producers through participation in

farm work. This reluctance reflects deeper structural barriers. The difficulties in this area can be attributed to the low level of social capital. The socialist period weakened community bonds, and attempts to collectivize agriculture along the Soviet model made Poles skeptical toward cooperative forms of organization. The low level of social capital also translates into a lack of shared responsibility among consumers and producers for local development and for local food and economic security.
- The Principle of Proximity: This principle presents both opportunities and challenges for CSA development. On one hand, it is reinforced by ongoing urban sprawl, which shortens the distance between potential consumers (urban middle-class or professionals) and farms. On the other hand, assuming that most potential consumers are based in large cities, this principle creates barriers to CSA development in more remote, peripheral rural areas.
- The Principle of Diversity: This is a key principle, particularly from the consumer perspective. The primary motivator for consumer participation in CSA is access to high-quality food. For farmers, diversity can serve as a strategy of diversifying agricultural operations. This principle is also essential for fulfilling the CSA postulate of sustainable rural development. Diversity likewise distinguishes CSA from industrial agriculture, where—particularly in the case of large-scale farms—production is dominated by monocultures, such as oilseed crops or cereals.

Based on the above analysis, a CSA model tailored to the Polish context may emerge, characterized by the following features:

- Involvement of medium-sized farms (11–50 ha), located within approximately 70 kilometers of the delivery point. This conclusion is based on several factors:
    - A substantial number of such farms exists; in 2020, there were approximately 329,000 such farms in Poland, accounting for 22.9% of all farms in the country;
    - These farms are capable of engaging in diversified agricultural production;
    - Their size allows for the allocation of part of the land to organic or traditional farming;
    - Transport costs and delivery times do not constitute barriers to fulfilling customer orders.
- Participation of consumers residing in cities with populations over 100,000 or in adjacent commuter municipalities (urban suburbs). Consumers will not be required to make advance payments. Instead, contracts will guarantee them priority access to agricultural products. Consumers will not be regularly involved in farm labor. Any participation will be incidental and linked to educational purposes, such as

learning about ecological or traditional farming practices and animal husbandry.

- Involvement of municipalities or non-governmental organizations (NGOs). These entities should serve as guarantors of the model's stability. They are expected to establish the legal and economic framework within which CSA can operate. Their involvement should help overcome the currently low level of social capital. These entities would be responsible for initiating CSA partnerships and overseeing their functioning. Additionally, they should provide facilities for the local collection and distribution of agricultural products.

In conclusion, the new CSA model should foster stable and transparent relationships between farmers and consumers, while also ensuring the active involvement of local government units or non-governmental organizations. These entities should provide the necessary economic and legal framework to reduce transaction costs and increase the overall attractiveness of the model.

In the evolving CSA model, the business and environmental dimensions are likely to become predominant, whereas the principle of solidarity will be significantly reduced. This shift represents a marked departure from the original foundations of CSA and from the models implemented in Western European countries. Nevertheless, in post-socialist states—particularly in Poland—such a liberalized approach appears necessary if CSA is to move beyond its current market niche and become a genuine alternative to industrial agriculture. Although CSA currently does not have the capacity to radically transform the dominant paradigm of agricultural development, it may contribute to the long-term evolution of the system.

A transition toward a more business-oriented model will require farmers to broaden the range of products they offer and to demonstrate greater professionalism in production planning and management. It also appears advisable to promote networking among CSA initiatives and to make use of synergies between farms. Such an approach could expand the available product range without increasing labor or land productivity and without lengthening supply chains.

It may also be worth considering the introduction of a formal legal structure for CSA—for example, in the form of a cooperative. Its members could include farmers and consumers and, optionally, non-governmental organizations and municipal authorities. As demonstrated by the experience of energy cooperatives, granting a clear legal form to such initiatives enhances their durability and operational effectiveness.

General Conclusion

Based on the research conducted, it can be concluded that CSA is one of the alternative food system models that promotes direct relationships between producers and consumers, supports the local economy, minimizes environmental impact, and addresses challenges related to food security and sustainable development.

A review of the literature and the findings from empirical studies lead to the following overarching conclusions:

- The CSA model represents an innovative approach to organizing agricultural production, offering tangible benefits to both farmers and consumers. However, as experiences from other countries—including the United States—have shown, effective CSA cooperation requires a long-term perspective and must take into account the investment needs of farms.
- CSA constitutes a form of constructive innovation—aligned with the paradigm of sustainable development. It positively influences the economic, social, and environmental dimensions of rural development. Furthermore, it helps strike a balance between increasing productivity and preserving the natural environment and agricultural heritage.
- The literature suggests that Poland has considerable potential for CSA development, partly due to the so-called backwardness rent, arising from a fragmented agrarian structure that has slowed the expansion of industrial agriculture. A similar situation can also be observed in other countries of the region, particularly Romania.
- Unfortunately, the development of CSA in Poland is hindered by several barriers—most notably a lack of trust and a reluctance to cooperate, both of which stem from a persistently low level of social capital. This low social capital is largely a legacy of historical processes, particularly the experience of a centrally planned economy. This issue is also visible in other Central and Eastern European countries. It may therefore be argued that it is characteristic of post-communist economies.

DOI: 10.4324/9781003745198-7

- The persistently low level of social capital indicates that the widespread implementation of a Western-style CSA model in Poland is unlikely. Without adapting the CSA model to the specific conditions of Central and Eastern European countries, it will remain a niche solution.
- These deficits in human and social capital should be offset by the involvement of the public sector in developing and supporting this model, particularly through local government authorities and non-governmental organizations (NGOs). However, current engagement remains low, and there is a noticeable lack of understanding of the essence of the CSA model. As a result, local governments and agricultural advisory centers have not developed institutional mechanisms to support the establishment and functioning of CSA initiatives.
- In practice, CSA in Poland includes not only small farms (up to 10 ha) but also medium- and large-scale farms that treat the model as an additional source of income. This indicates a need to reconsider existing assumptions about the applicability of CSA under Polish conditions. It may also suggest that, in the future, CSA will evolve into a more business-oriented model, gradually losing its social and solidarity-based dimension.
- Although consumers express interest in CSA, they primarily see it as a way to access organic and traditional food products. The principle of solidarity, while occasionally mentioned, tends to be more declarative than practical—there is little evidence that consumers are genuinely willing to share production risks with farmers.
- A new CSA model for Poland—potentially considered a transitional model—should move away from the Principle of Solidarity and instead emphasize Partnership, Proximity, and Diversity. This model should include local governments and non-governmental organizations as key facilitators.

The findings highlight the potential of CSA as a model of cooperation between farmers and consumers, particularly in meeting the needs of niche markets. Strengthening social capital, expanding educational initiatives, and ensuring stable institutional support will be crucial for its successful development in Poland. Importantly, the results confirm the research questions posed at the outset.

CSAs in Central and Eastern Europe differ from those in Western Europe due to distinct historical, economic, and social conditions. At the current stage of socio-economic development in Central and Eastern European countries, transferring solutions from the "old EU" lacks justification. A rigid adherence to the original principles would prevent CSA from expanding beyond a niche. Owing to the low level of social capital—an enduring legacy of communism—building long-term relationships between producers and consumers remains particularly challenging. Although CSA in the region has notable development potential, adapting the model to local conditions, including a reconsideration of the principle of solidarity, may be necessary.

Overall, the study demonstrates that the long-term future of CSA in Poland will depend on strengthening social capital, implementing systematic educational efforts, and providing coherent and durable institutional support, without which the model will remain marginal.

Bibliography

Actu Environnement, 2017. https://www.actu-environnement.com/ae/news/circuits-courts-proximite-alimentation-avis-Ademe-29282.php4

Adam K. L., 2006. *Community Supported Agriculture*. National Sustainable Agriculture Information Service. National Center for Appropriate Technology, Butte, MT.

Adamowicz M., Dresler E., 2006. Zrównoważony rozwój obszarów wiejskich na przykładzie wybranych gmin woj. lubelskiego. *Zeszyty Naukowe Akademii Rolniczej we Wrocławiu. Rolnictwo* LXXXVII, Nr 540.

Adger W. N., 2000. Social and ecological resilience: Are they related?. *Progress in Human Geography*, 24(3), 347–364.

Allen T., Gabe T., McConnon J., 2006. *The Economic Contribution of Agri-Tourism to the Maine Economy*. University of Maine Department of Resource Economics and Policy, Orono.

Altieri M. A., Nicholls C. I., 2017. The Adaptation and Mitigation Potential of Traditional Agriculture in a Changing Climate. *Climatic Change*, 140, 33–45. https://doi.org/10.1007/s10584-013-0909-y

ARC, 2020. https://www.arc2020.eu/agroecology/czech-republic/

Arnstein S. R., 1969. A Ladder of Citizen Participation. *Journal of the American Institute of Planners*, 35(4), 216–224.

Bakos I. M., Khademi-Vidra A., 2019. Empirical Experiences of the Hungarian Alternative Food Buying Communities. *Deturope – The Central European Journal of Regional Development and Tourism*, 11(1), 55–73.

Balázs B., 2012. Local Food System Development in Hungary. *International Journal of Sociology of Agriculture and Food*, 19(3), 403–421.

Balázs B., Pataki G., Lazányi O., 2016. Prospects for the Future: Community Supported Agriculture in Hungary. *Futures*, 83, 100–111.

Banks J., Bristow G., 1999. Developing Quality in Agro-Food Supply Chains: A Welsh Perspective. *International Planning Studies*, 4, 317–331.

Bashford J., Cross K., Eichinger W., Georgakakis A., Iserte M., Kern F., Lešinský D., Pabst S., Parot J., Perényi Z., Valeška J., Wendland M., 2015. *Guide Européen sur l'Agriculture soutenue par la communauté*. Community Supported Agriculture for Europe CSA, Vienna.

Batowski M., 2009. *Gospodarka socjalistyczna w Polsce. Geneza – rozwój – upadek*, Wydawnictwo Naukowe PWN, Warszawa.

Benedek Z., Fertő I., Molnár A., 2018. Off to Market But Which One? Understanding the Participation of Small-Scale Farmers in Short Food Supply Chains – A Hungarian Case Study. *Agriculture and Human Values*, 35(2), 383–398. https://doi.org/10.1007/s10460-017-9834-4

Bernard K., Bonein A., Bougherara D., 2020. Consumer Inequality Aversion and Risk Preferences in Community Supported Agriculture. *Ecological Economics*, 175, 106684.

Bilewicz A., Śpiewak R., 2019. Beyond the "Northern" and "Southern" divide: Food and space in Polish consumer cooperatives. *East European Politics and Societies and Cultures*, 33(3), 579–602.

Binswanger H. C., 2012. *The Growth Spiral: Money, Energy, and Imagination in the Dynamics of the Market Process*. Springer, St. Gallen.

Bîrhală B., Möllers J., 2012. *Community Supported Agriculture in Romania: Is it driven by economy or solidarity?*. IAMO, Leibniz Institute of Agricultural Development in Transition Economies. https://www.econstor.eu/bitstream/10419/92936/1/778143678.pdf

Bîrhală B., Möllers J., 2014. Community Supported Agriculture: Is It Driven by Economy or Solidarity?. *Discussion Paper* No. 144, Halle (Saale). http://hdl.handle.net/10419/92936

Birtalan I. L., Bartha A., Neulinger Á., Bárdos G., Oláh A., Rácz J., Rigó A., 2020. Community Supported Agriculture as a Driver of Food-Related Well-Being. *Sustainability*, 12, 4516. https://doi.org/10.3390/su12114516

Blättel-Mink B., Boddenberg M., Gunkel L., Schmitz S., Vaessen F., 2017. Beyond the Market—New Practices of Supply in Times of Crisis: The Example Community-Supported Agriculture. *International Journal of Consumer Studies*, 41(4), 415–421. https://doi.org/10.1111/ijcs.12351

Błażejowska M., Gostomczyk W., 2018. Warunki tworzenia i stan rozwoju spółdzielni i klastrów energetycznych w Polsce na tle doświadczeń niemieckich. *Zeszyty Naukowe SGGW w Warszawie – Problemy Rolnictwa Swiatowego*, 18(33).

Bloemmen M., Bobulescu R., Le N. T., Vitari C., 2015. Microeconomic Degrowth: The Case of Community Supported Agriculture. *Ecological Economics*, 112, 110–115. https://doi.org/10.1016/j.ecolecon.2015.02.013

Bobulescu R., Tuyen Le N., Vitari C., Whittingham E., 2018. Socio-Economic and Ecological Transition in Community Supported Agriculture: From the "transitional" to the "ideal" CSA. *International Journal of Agricultural Resources, Governance and Ecology*, 14(2), 122–137.

Bonfert B., 2022. Community-Supported Agriculture Networks in Wales and Central Germany: Scaling Up, Out, and Deep through Local Collaboration. *Sustainability*, 14, 7419.

Bourdieu P., 1986. The Forms of Capital. In: *Handbook of Theory and Research for the Sociology of Education*, edited by Richardson J. G., Greenwood Press, New York, 241–258.

Brown C., Miller S., 2008. The Impacts of Local Markets: A Review of Research on Farmers Markets and Community Supported Agriculture (CSA). *American Journal of Agricultural Economics*, 90, 1296–1302. https://doi.org/10.1111/j.1467-8276.2008.01220.x

Brundtland G., 1987. *Our Common Future*. Report of the World Commission on Environment and Development, Oxford University Press, Oxford–New York.

Brunori G., Rossi A., Guidi F., 2012. On the New Social Relations around and beyond Food. Analysing Consumers' Role in the Gruppi di Acquisto Solidale (Solidarity Purchasing Groups). *Sociologia Ruralis*, 52(1), 1–30.

Burja V., Tamas-Szora A., Dobra I. B., 2020. Land Concentration, Land Grabbing and Sustainable Development of Agriculture in Romania. *Sustainability*, 12, 2137. https://doi.org/10.3390/su12052137

Carlson L. A., Bitsch V., 2019. Applicability of Transaction Cost Economics to Understanding Organizational Structures in Solidarity-Based Food Systems in Germany. *Sustainability*, 11, 1095. https://doi.org/10.3390/su11041095

Chaturvedi A., Armstrong B., Chaturvedi R., 2014. Securing the Food Supply Chain: Understanding Complex Interdependence through Agent-Based Simulation. *Health Technology*, 4(2), 159–169.

Chen W., 2013. Perceived Value of a Community Supported Agriculture (CSA) Working Share. The construct and its dimensions. *Appetite*, 62, 37–49.

Chicoma J. L., 2024. *Supporting Food Systems Transformation Towards Sustainability and Resilience*. United Nations Development Programme, New York. https://www.undp.org/sites/g/files/zskgke326/files/2024-09/undp-supporting-food-systems-transformation-towards-sustainability-and-resilience.pdf

Chiffoleau Y., 2019. *Les circuits courts alimentaires. Entre marché et innovation sociale*. Éditions érès, Toulouse.

Chiffoleau Y., Wallet F., Prosperi P., Collewet F., Morizot-Braud F., 2025. Short Food Supply Chains, an Alternative Economy for Sustainable Food Systems: A Mixed Approach Over Time. *Agricultural and Food Economics*, 13, 66. https://doi.org/10.1186/s40100-025-00403-4

Coleman J., 1990. *Foundations of Social Theory*. Belknap Press of Harvard University Press, Cambridge, 302.

COM, 2019. 640 final, *The European Green Deal*, European Commission.

COM, 2020a. 380 final. *Communication from the Commission to the European Parliament, the Council, the European Economic and Social Committee and the Committee of the Regions*. EU Biodiversity Strategy for 2030 Bringing nature back into our lives.

COM, 2020b. 381 final. *Communication from the Commission to the European Parliament, the Council, the European Economic and Social Committee and the Committee of the Regions*. A Farm to Fork Strategy for a fair, healthy and environmentally-friendly food system.

COM, 2020. Reinforcing Europe's resilience: halting biodiversity loss and building a healthy and sustainable food system. https://ec.europa.eu/commission/presscorner/detail/en/ip_20_884

Cone C. A., Myhre A., 2000. Community-Supported Agriculture: A Sustainable Alternative to Industrial Agriculture? *Human Organization*, 59(2), 187–197.

Conner D. S., 2003. *Community Supported Agriculture Pricing and Promotion Strategies: Lessons from Two Ithaca, NY Area Farms*. Department of Applied Economics and Management, College of Agriculture and Life Sciences, Cornell University, Ithaca, NY.

CRIES, 2011. *Parteneriat Local pentru o Economie Sustenabilă*. Timişoara.

Cristiano S., 2021. Organic Vegetables from Community-Supported Agriculture in Italy: Emergy Assessment and Potential for Sustainable, Just, and Resilient Urban-Rural Local Food Production. *Journal of Cleaner Production*, 292, 126015.

Cristiano S., Auriemma M., Cacciari P., Cervesato M., Maffeo D., Malgaretto P., et al., 2020. Nourishing Self-Planned Socioecological Transformations: Glocal Community Supported Agriculture in Veneto, Italy. In *Food for Degrowth*, edited by Nelson A., Edwards F., Routledge, London, 90–99.

Czyżewski A., 2015. Teoriopoznawcze przesłanki rozwoju rolnictwa rodzinnego. In *Ekonomiczne mechanizmy wspierania i ochrony rolnictwa rodzinnego w Polsce i innych państwach Unii Europejskiej*, edited by Chlebicka A., FAPA i MRiRW, Warszawa.

Daly H., 1996. *Beyond Growth: The Economics of Sustainable Development*. Beacon Press., Boston.

Dankar H., 2021. *Community-supported Agriculture (CSA)*. NC State Extension, NC State University, https://localfood.ces.ncsu.edu/local-food-marketing-markets/local-food-csa/ (accessed 13.6.23).

Darnhofer I., Fairweather J., Moller H., 2010, Assessing a Farm's Sustainability: Insights from Resilience Thinking. *International Journal of Agricultural Sustainability*, 8(3), 186–198. https://doi.org/10.3763/ijas.2010.0480

Dekret z dnia 6 września 1946 r. o ustroju rolnym i osadnictwie na Ziemiach Odzyskanych i obszarze byłego Wolnego Miasta Gdańska. *Dz. U.*, 1946 Nr 49, poz. 279.

DeLind L. B., 1999, Close Encounters with a CSA: the Reflections of a Bruised and Somewhat Wiser Anthropologist. *Agriculture and Human Values*, 16(3), 3–9.

DeMuth S., 1993, *Community Supported Agriculture (CSA): An Annotated Bibliography and Resource Guide*, National Agricultural Library, Beltsville.

Devere C., Lamine C., 2010, Les systèmes agroalimentaires alternatifs. Une revue de travaux anglophones en sciences sociales. *Économie rurale*, 317, 57–73.

Diekmann M., 2020, Community Supported Agriculture – Innovative Nischenstrategie für Landwirtschaftliche Betriebe?. *Berichte ber Landwirtschaft*, 98, 1–21.

Drucker P.F., 2006, *Innovation and Entrepreneurship*. Harper Business, New York.

Egartner S., Niedermayr J., Ayrault J., 2020, Development of Community Supported Agriculture in Austria and France. *Journal of the Austrian Society of Agricultural Economics*, 29, 28.

Egli L., Rüschhoff J., Priess J., 2023, A Systematic Review of the Ecological, Social and Economic Sustainability Effects of Community-Supported Agriculture. *Frontiers in Sustainable Food Systems*, 7, 1–11.
Ellison G., Pino N.W., 2012, Development Paradigms, Neoliberalism and Human Security — Western Style. In *Globalization, Police Reform and Development*, edited by Ellison G, Pino N . Transnational Crime, Crime Control and Security. Palgrave Macmillan, London.
EMAC Annual Conference, European Marketing Academy (EMAC). BEL, June 2014, Valence, Spain.
Enthoven L., van den Broeck G., 2021, Local Food Systems: Reviewing Two Decades of Research. *Agricultural Systems*, 193, 103226.
Ertmańska K., 2015, Community Supported Agriculture (CSA) as a form of Sustainable Consumption. *Oeconomia*, 14(2), 51–59.
Espelt R., 2020, Agroecology Prosumption: The Role of CSA Networks. *Journal of Rural Studies*, 79, 269–275.
European Commission, 2024, Proofing the EU Food Supply Chain Against Crises: New Set of Recommendations Published. https://agriculture.ec.europa.eu/news/proofing-eu-food-supply-chain-against-crises-new-set-recommendations-published-2024-07-23_en
Fagerberg J., 2005, Innovation a Guide to the Literature. In *The Oxford Handbook of Innovation*, edited by Fagerberg J., Mowery D.C., Nelson R.R., Oxford University Press, New York, 155.
FAO, 2021, *Small Family Farmers Produce a Third of the World's Food*. FAO Research Study. Food and Agriculture Organization of the United Nations, Rome.
Farnsworth R.L., Thompson S.R., Drury K.A., Warner R.E., 1996, Community Supported Agriculture: Filling a Market Niche. *Journal of Food Distribution Research*, 27(1), 90–98.
Feagan R., Henderson A., 2009, Devon Acres CSA: Local Struggles in a Global Food System. *Agriculture and Human Values*, 26, 203–217.
FER, https://fer.org.pl/
Fiedor B., 2002, Teoria trwałego rozwoju jako krytyka neoklasycznej ekonomii środowiska i wzrostu gospodarczego. In *Podstawy ekonomii środowiska i zasobów naturalnych*, , edited by Fiedor B., Wydawnictwo C.H. Beck, Warszawa, 225–255.
Fiedor B., 2011, Trwały rozwój a koncepcja społecznej gospodarki rynkowej. *Prace naukowe Uniwersytetu Ekonomicznego we Wrocławiu*, 225, 13–14.
Fletcher R., Rammelt C., 2017, Decoupling: A Key Fantasy of the Post-2015 Sustainable Development Agenda. *Globalizations*, 14(3), 450–467.
Flora C.B., Bregendahl C., 2012, Collaborative Community-Supported Agriculture: Balancing Community Capitals for Producers and Consumers. *International Journal of Sociology of Agriculture and Food*, 19, 329–346. https://doi.org/10.48416/ijsaf.v19i3.208
Fomina Y., Glińska-Neweś A., Ignasiak-Szulc A., 2022, Community Supported Agriculture: Setting the Research Agenda Through a Bibliometric Analysis. *Journal of Rural Studies*, 92, 294–305.

Fonte M., 2013, Food Consumption as Social Practice: Solidarity Purchasing Groups in Rome, Italy. *Journal of Rural Studies*, 32, 230–239.

Freeman C., 1982, *The Economics of Industrial Innovation*. F. Pinter, London, 7.

Fukuyama F., 1995, *Trust: The Social Virtues and the Creation of Prosperity*. The Free Press, New York.

Fukuyama F., 2001, Social Capital, Civil Society and Development. *Third World Quarterly*, 22(1), 7.

Gallaud D., Laperche B., 2016, *Economie circulaire et dévelopment durable, écologie industrielle et circuits courts*. ISTE Editions, London.

Galt R.E., 2013, The Moral Economy Is a Double-edged Sword: Explaining Farmers' Earnings and Self-exploitation in Community-Supported Agriculture. *Economic Geography*, 89(4), 341–365. https://doi.org/10.1111/ecge.12015

Galt R.E., Bradley K., Christensen L., Fake C., Munden-Dixon K., Simpson N., Surl R., Van Soelen Kim J., 2017, What Difference Does Income Make for Community Supported Agriculture (CSA) Members in California? Comparing Lower-Income and Higher-Income Households. *Agriculture and Human Values*, 34, 435–452.

Galt R.E., Bradley K., Christensen L., van Soelen Kim J., Lobo R., 2016, Eroding the Community in Community Supported Agriculture (CSA): Competition's Effects in Alternative Food Networks in California. *Sociologia Ruralis*, 56(4), 491–512. https://doi.org/10.1111/soru.12102

Geels F., 2005, *Technological Transitions and System Innovations: A Co-Evolutionary and Socio-Technical Analysis*. Edward Elgar Publishing, Cheltenham, Northampton.

Gereffi G., Humphrey J., Sturgeon T., 2005, The Governance of Global Value Chain. *Review of International Political Economy*, 12(1), 78–104.

Gibson-Graham J.K., 2006, *A Postcapitalist Politics*. University of Minnesota Press, Minneapolis, London.

Gliessman S., 2016, Transforming Food Systems with Agroecology. *Agroecology and Sustainable Food Systems*, 40(3), 187–189. https://doi.org/10.1080/21683565.2015.1130765

Goodman D., Goodman M., 2009, Alternative Food Networks. In *International Encyclopedia of Human Geography*, edited by Kitchin R., Thrift N., Elsevier, Amsterdam.

Góral J., Rembisz W., 2017, Wynagrodzenia i wydajność pracy w rolnictwie polskim na tle pozostałych krajów Unii Europejskiej. *Zeszyty Naukowe SGGW – Ekonomika i Organizacja Gospodarki Zywnościowej*, 119, 119–139.

Gorman R., 2018, Human-Livestock Relationships and Community Supported Agriculture (CSA) in the UK. *Journal of Rural Studies*, 61, 175–183.

Dz U. 1944 Nr 4, poz. 17. W dekrecie doprecyzowano m.in., że za uzyskaną ziemię chłopi będą płacić równowartość przeciętnych rocznych zbiorów, ale należność została rozłożona na 10–20 lat.

Gorzelak E., 2010, *Polskie rolnictwo w XX wieku. Produkcja i ludność*. Prace i Materiały Instytutu Rozwoju Gospodarczego SGH, Warszawa, 19.
Goszczyński W., 2019, In Search of the Vocabulary for Eastern European Food Studies. Conceptual Remarks After the Workshop: Alternative Food Supply Networks in Central and Eastern Europe. *Eastern European Countryside*, 25, 273–286. https://doi.org/10.12775/EEC.2019.014
Goszczyński W., Knieć W., 2011, Strengthening Alternative Agro-food Networks in the Eastern European Countryside. *Eastern European Countryside*, 17(1), 5–20. https://doi.org/10.2478/v10130-011-0001-4
Goszczyński W., Wróblewski M., Wójtowicz A., 2018, Jakość w polskich alternatywnych sieciach żywnościowych. Analiza praktyk społecznych. *Studia Socjologiczne*, 1(228), 146.
Gradka I., Jstrzębska E., Sałata B., Żakowska-Biemas S., 2023, *Przetwórstwo i sprzedaż produktów ekologicznych*. Centrum Doradztwa Rolniczego w Brwinowie Oddział w Radomiu, Radom.
Groh T., Mc Fadden S., 1997, *Farms of Tommorrow Revisited. Community supported farms, Farm supported communities*. Biodynamic Farming and Gardening Association, Kimberton.
Groh T., McFadden S., 1990, *Farms of Tomorrow: Community Supported Farms – Farm Supported Communities*. Biodynamic Farming and Gardening Association, Kimberton.
Gruber S., 2020, *Bewältigungsstrategien alternativen Wirtschaftens: Wertrationalität und soziale Einbettung am Beispiel solidarischer Landwirtschaft*. Nomos, Baden-Baden. https://doi.org/10.5771/97837 48909194
Haack M., Engelhardt H., Gascoigne C., Schrode A., Fienitz M., Meyer-Ohlendorf L., et al., 2020, *Nischen des Ernährungssystems: Bewertung des Nachhaltigkeits- und Transformationspotenzials innovativer Nischen des Ernährungssystems in Deutschland*. Umweltbundesamt, Dessau-Roßlau.
Harman A. J., 1971, *The International Computer Industry. Innovation and Comparative Advantage*. Harvard University Press, Cambrige, MA, 151–169.
Harmon H. A., 2014, Community Supported Agriculture: A Conceptual Model of Health Implications. *Austin J. Nutr. Food Sci.*, 2, 1024.
Harvey D., 2005, *A Brief History of Neoliberalism*. Oxford University Press, Oxford.
Havadi-Nagy K. X., 2021, Alternative Food Networks in Romania – Effective Instrument for Rural Development?. *Journal of Settlements and Spatial Planning*, Special Issue No. 8, 15–27. https://doi.org/10.24193/JSSPSI.2021.8.03
Hayek Von F. A., 2007, *Konstytucja wolności*, Wydawnictwo Naukowe PWN, Warszawa, 65.
Henderson E., Van En R., 2007, *Sharing the Harvest: A Citizen's Guide to Community Supported Agriculture*, Chelsea Green Publishing, Revised & enlarged edition, Vermont.
Hinrichs C., 2000, Embeddedness and Local Food Systems: Notes On Two Types of Direct Agricultural Market. *Journal of Rural Studies*, 16(3), 295–303.

Howard M. M., 2003, *The Weakness of Civil Society in Post-Communist Europe*. Cambridge University Press, Cambridge.

Ilbery B., Maye D., 2005a, Alternative (Shorter) Food Supply Chains and Specialist Livestock Products in the Scottish–English Borders. *Environment and Planning A*, 37, 823–844.

Ilbery B., Maye D., 2005b, Food Supply Chains and Sustainability: Evidence from Specialist Food Producers in the Scottish/English Borders. *Land Use Policy*, 22, 331–344.

IPES-Food, 2024, *Food from Somewhere: Building Food Security and Resilience Through Territorial Markets*. International Panel of Experts on Sustainable Food Systems. https://www.ideassonline.org/public/pdf/IPES-FoodReport2024-ENG.pdf

Jackson T., 2009, *Prosperity without Growth? The Transition to a Sustainable Economy*. Sustainable Development Commission. https://www.growthintransition.eu/wp-content/uploads/prosperity_without_growth_report.pdf

Jehlička P., Daněk P., Vávra J., 2019, Rethinking Resilience: Home Gardening, Food Sharing and Everyday Resistance. *Canadian Journal of Development Studies / Revue canadienne d'études du développement*, 40(4), 511–527. https://doi.org/10.1080/02255189.2018.1498325

Jehlička P., Pungas L., 2022, Everyday Sustainability and Post-Socialist Households. *Sustainability*, 14(2), 789.

Jehlička P., Smith J., 2011, An Unsustainable State: Contrasting Food Practices and State Policies in the Czech Republic. *Geoforum*, 42(3), 362–372. https://doi.org/10.1016/j.geoforum.2011.01.005

Kachniarz M., Raftowicz M., 2020, The Institutional Model of Local Food Market Organization with the Participation of Local Government. In *KNOWCON 2020. Knowledge on Economics and Management. Conference Proceedings*, edited by Slavíčková P., Stoklasa J., Palacký University Olomouc, Olomouc, 59–65.

Kacz K., Hegyi J., Gombkötő N., 2019, Characteristics of Community Supported Agriculture in the Western Transdanubia Region. *Deturope – The Central European Journal of Regional Development and Tourism*, 11(1), 42–54.

Kaliński J., 2005, *Historia gospodarcza Polski Ludowej*, Wydawnictwo: Uniwersytet w Białymstoku, Białystok.

Kaliński J., Landau Z., 2003, *Gospodarka Polski w XX wieku*, Polskie Wydawnictwo Ekonomiczne, Warszawa.

Kallis G., 2018, *Degrowth*, Agenda Publishing, Newcastle-upon-Tyne.

Kamerschen D. R., McKenzie R. B., Nardinelli C., 1991, *Ekonomia*, Wydawnictwo: Fundacja Gospodarcza NSZZ "Solidarność", Gdańsk.

Kemp K., Insch A., Holdsworth D. K., Knight J. G., 2010, Food Miles: Do UK Consumers Actually Care?. *Food Policy*, 35(6), 504–513.

Khmara Y., Kronenberg J., 2020, Degrowth in the Context of Sustainability Transitions: In Search of a Common Ground. *Journal of Cleaner Production*, 267, 122072.

Koch M., Buch-Hansen H., 2020, In Search of a Political Economy of the Postgrowth Era. *Globalizations*.

Kociszewski J., 1999, *Proces integracji ziem zachodnich i północnych z Polską*, Wydawnictwo Akademii Ekonomicznej we Wrocławiu, Wrocław.
Kociszewski K., 2013, *Ekologizacja polskiego rolnictwa a jego zrównoważony rozwój w warunkach członkostwa w Unii Europejskiej*, Wydawnictwo Uniwersytetu Ekonomicznego we Wrocławiu, Wrocław, 29.
Kolodinsky J. M., Pelch L. L., 1997, Factors Influencing the Decision to Join a Community Supported Agriculture (CSA) Farm. *Journal of Sustainable Agriculture*, 10, 129–141.
Kołodko G., 2014, Nowy pragmatyzm, czyli ekonomia I polityka dla przyszłości. *Ekonomista*, 214/3, 161–180.
Kołodziejczyk D., Gospodarowicz M., Lidke D., Miś T., Raftowicz-Filipkiewicz M., Łabędzki H., Struś M., 2015, *Wieloaspektowe ujęcie problemu spójności obszarów wiejskich na poziomie lokalnym i regionalnym w dotychczas realizowanych i planowanych strategiach rozwoju*, Instytut Ekonomiki Rolnictwa i Gospodarki Żywnościowej Państwowy Instytut Badawczy, Warsaw, 92.
Kondo C., 2021, Re-energizing Japan's Movement: Understanding Intergenerational Transitions of Diverse Economies. *Journal of Agriculture, Food Systems, and Community Development*, 10(4).
Kondoh K., 2015, The Alternative Food Movement in Japan: Challenges, Limits, and Resilience of the Teikei System. *Agriculture and Human Values*, 32, 143–153.
Kotler P., 2017, *Principles of Marketing*, 17th Edition, Pearson.
Kovách I., Megyesi B., 2024, Agricultural Involvement in Hungarian Society Since EU Accession. *Wieś i Rolnictwo*, 4(205), 69–87. https://doi.org/10.53098/wir.2024.4.205/03
Kraiß K., van Elsen T., 2009, Landwirtschaftliche Wirtschaftsgemeinschaften (Community Supported Agriculture, CSA) – ein Weg zur Revitalisierung des ländlichen Raumes?. In *Nachhaltige Entwicklung ländlicher Räume*, , edited by Friedel R., Spindler E. A., VS Verlag, Wiesbaden, 183–194.
KSOW, https://ksow.pl/baza-danych/lgd-2014-2020
Kutkowska B. (Eds.), 2018, *Rolnictwo Wspierane Społecznie – badanie szans rozwoju małych gospodarstw rolnych na Dolnym Śląsku w aspekcie produkcji żywności wysokiej jakości*, Wydawnictwo Uniwersytetu Przyrodniczego we Wrocławiu, Wrocław.
Kutkowska B. (red.), 2017, *Rolnictwo Wspierane Społecznie – zmniejszenie barier wejścia na rynek dla dolnośląskich produktów żywności wysokiej jakości*, Wydawnictwo Uniwersytetu Przyrodniczego we Wrocławiu, Wrocław.
Łabędzki H., Struś M., 2014, Społeczne determinanty rozwoju przygranicznych obszarów wiejskich w południowo-zachodniej Polsce. *Prace Naukowe Uniwersytetu Ekonomicznego we Wrocławiu*, 341, 136–145.
Lagane J., 2015, When Students Run AMAPs: Towards a French Model of CSA. *Agriculture and Human Values*, 32, 133–141
Lang K. G., 2010, The Changing Face of Community-Supported Agriculture. *The Journal of Culture and Agriculture*, 32(1), 17–26.

Latouche S., 2010, *Le pari de la décroissance*. Pluriel, Paris.
Laville J.-L., 2010, The Solidarity Economy: An International Movement. *RCCS Annual Review*, 2. https://journals.openedition.org/rccsar/202
le Vay C., 1983, Agricultural Co-operative Theory. *The Review Journal of Agricultural Economics*, 34(1), 1–12.
Le Velly R., Dufeu I., 2015, Alternative Food Networks as "Market Agencements": Exploring Their Multiple Hybridities. *Journal of Rural Studies*, 43, 173–182.
Légifrance, 2018, LOI n° 2018-938 du 30 octobre 2018 pour l'équilibre des relations commerciales dans le secteur agricole et alimentaire et une alimentation saine, durable et accessible à tous. https://www.legifrance.gouv.fr/jorf/id/JORFTEXT000037547946
Lewis W. W., 2005, *Potęga wydajności*. CeDeWu, Warszawa.
Lodkowska-Skoneczna A., Pyszkowski A., Szlachta J., 1996, Rozwój regionalny Polski 1990–1995 (skrót raportu diagnostycznego). *Gospodarka Narodowa*, 6, 7–8.
Mair S., Druckman A., Jackson T., 2020, A Tale of Two Utopias: Work in a Post-Growth World. *Ecological Economics*, 173, 106653.
Malak Rawlikowska A., Majewski E., Was A., Borgen S. O., Csillag P., Donati M., Freeman R., Hoang V., Lecoeur J.-L., Mancini M., Nguyen A., Monia S., Tocco B., Torok A., Veneziani M., Vittersø G., Wavresky P., 2019, Measuring the Economic, Environmental, and Social Sustainability of Short Food Supply Chains. *Sustainability*, 11, 4004.
Małysz J., 2002, Rozwój agrobiznesu a procesy integracyjne, *Wieś i Rolnictwo*, 1(114), 25–30.
Manifest PKWN, *Dokumenty programowe polskiego ruchu robotniczego 1878–1984*, red. N. Kołomejczyk, B. Syzdek, Książka i Wiedza, Warszawa, 1986, 348.
Mansfield E., 1968, *Industrial Research and Technological Innovation*, W. W. Norton, New York, 83.
Mansura M., 2024, The Fragile Link: Supply Chain Disruptions and Global Food Security. *Food Science and Technology*, 38(3), 36–39.
Marquat C., Diemer A., 2016, Les AMAP, une émanation de l'économie sociale et solidaire. *Revue Francophone du Développement Durable*, 7, mars, 1–21.
Martinez S., Hand M., Da Pra M., Pollack S., Ralston K., Smith T., Vodel S., Clark S., Lohr L., Low S., Newman C., 2010, *Local Food System, Concepts, Impacts, and Issues*. United States Department of Agriculture, Economic Research Service, Washington DC, 97.
Mastronardi L., Marino D., Giaccio V., et al., 2019, Analyzing Alternative Food Networks Sustainability in Italy: A Proposal for An Assessment Framework. *Agric Econ*, 7, 21.
Matysiak A., Struś M., 2015, Paradygmat rozwoju zrównoważonego. *Studia Ekonomiczne. Zeszyty Naukowe Uniwersytetu Ekonomicznego w Katowicach*, 213, 11–21.
Matysiak A., 1999, *Źródła kapitału społecznego*, Wydawnictwo Akademii Ekonomicznej we Wrocławiu, Wrocław, 61–62.

Matysiak A., Raftowicz M., 2019, Poliformizm systemu gospodarczego. *Studia Ekonomiczne. Zeszyty Naukowe Uniwersytetu Ekonomicznego w Katowicach*, 378, 7–18.

Matysiak A., Raftowicz-Filipkiewicz M., 2017, Wpływ procesów endogenicznych na rozwój zrównoważony [The influence of the endogenous processes on the sustainable development]. *Studia Ekonomiczne. Zeszyty Naukowe Uniwersytetu Ekonomicznego w Katowicach*, 311, 37–47.

Matzembacher D. E., Meira F. B., 2019, Sustainability as Business Strategy in Community Supported Agriculture. *BFJ*, 121, 616–632.

Meadows D. H., Meadows D. L., Randers J., Behrens W., 1972, *The Limits to Growth*. Universe Books, New York.

Medici M., Canavari M., Castellini A., 2023, An Analytical Framework to Measure the Social Return of Community-Supported Agriculture. *Agroecology and Sustainable Food Systems*, 47(9), 1319–1340.

Medici M., Canavari M., Castellini A., 2021, Exploring the Economic, Social, and Environmental Dimensions of Community-Supported Agriculture in Italy. *Journal of Cleaner Production*, 316, 128233.

Mędrzecki W., 2003, Chłopi. In *Społeczeństwo polskie w XX wieku*, red. J. Żarnowski, Wydawnictwo: Instytut Historii PAN, Warszawa.

Méndez V. E., Bacon C. M., Cohen R., 2013, Agroecology as a Transdisciplinary, Participatory, and Action-Oriented Approach. *Agroecology and Sustainable Food Systems*, 37, 3–18. https://doi.org/10.1080/10440046.2012.736926

Mert-Cakal T., Miele M., 2020, "Workable Utopias" for Social Change Through Inclusion and Empowerment? Community Supported Agriculture (CSA) in Wales as Social Innovation. *Agriculture and Human Values*, 1–20. https://doi.org/10.1007/s10460-020-10141-6

Meyer J. M., Hassler M., 2023, Re-Thinking Knowledge in Community-Supported Agriculture to Achieve Transformational Change towards Sustainability. *Sustainability*, 15, 13388.

Middendorf M., Rommel M., 2024, Understanding the Diversity of Community Supported Agriculture: A Transdisciplinary Framework with Empirical Evidence from Germany. *Frontiers in Sustainable Food Systems*, 8, 1–18.

Milford A. B., Keech D., Muiruri S. W., 2025, Community Supported Agriculture's Perceived Positive Impacts On Mental Health and Vegetable Consumption in Norway and the UK. *Agric Econ*, 13, 2.

Milford A. B., Reed M., 2024, Fresh Vegetables with a Story: Market Gardening for a Sustainable Food Transition. In *Plant-based food Consumption*, edited by Bertella G, Santini C. Elsevier, Amsterdam, 277–295.

Ministère de l'Agriculture et de la Souveraineté Alimentaire, https://agriculture.gouv.fr/pres-de-400-projets-alimentaires-territoriaux-pat-reconnus-par-le-ministere-au-1er-janvier-2023

Ministère de l'Agriculture et de la Souveraineté Alimentaire, https://agriculture.gouv.fr/quest-ce-quun-projet-alimentaire-territorial

Moellers J., Bîrhală B., 2014, Community Supported Agriculture: A Promising Pathway for Small Family Farms in Eastern Europe? A case study from Romania. *Landbauforschung*, 64(3/4), 139–150.

Moruzzi R., Sirieix L., 2014, Paradoxes of sustainable food and consumer coping strategies. *43. EMAC Annual Conference, European Marketing Academy (EMAC)*. BEL., Jun, Valencia, Spain. 9, ffhal-01594077.

Möllers J., Bîrhală B., 2014, Community Supported Agriculture: A Promising Pathway for Small Family Farms in Eastern Europe? A Case Study from Romania. *Landbauforschung – Applied Agricultural and Forestry Research*, 64(3/4), 139–150.

Möllers J., Traikova D., Bîrhală B. A. M., Wolz A., 2018, Why (not) Cooperate? A Cognitive Model of Farmers' Intention to Join Producer Groups in Romania. *Post-Communist Economies*, 30(1), 56–77.

Moruzzi R., Sirieix L., Paradoxes of Sustainable Food and Consumer Coping Strategies. 43.

Moulaert F., Nussbaumer J., 2005, The Social Region: Beyond the Territorial Dynamics of the Learning Economy. *European Urban and Regional Studies*, 12(1), 45–64.

Nascent, 2023, *Handbuch Solidarische Landwirtschaft*. https://www.solidarischelandwirtschaft.org/fileadmin/media/solidarischelandwirtschaft.org/Solawis-aufbauen/Handbuch/HandbuchSolidarische-Landwirtschaft-v1.2.pdf

Neilson D., 2019, Bringing in the 'Neoliberal Model of Development'. *Capital & Class*, 44(1), 85–108.

Netzwerk Solidarische Landwirtschaft e.V., 2022, Was ist Solidarische Landwirtschaft? https://www.solidarische-landwirtschaft.org/das-konzept/was-ist-solawi (accessed 25.01.2025).

North D., 1994, *Institutions, Institutional Change and Economic Performance*, Cambridge University Press, Cambridge.

North D. C., 2014, *Zrozumieć przemiany gospodarcze*, Oficyna Wolters Kluwer Business, Warszawa, 80.

Nost E., 2014, Scaling-up Local Foods: Commodity Practice in Community Supported Agriculture (CSA). *Journal of Rural Studies*, 34, 152–160.

Nowak A.,2022, Zróżnicowanie produktywności pracy w gospodarstwach rolnych w krajach członkowskich Unii Europejskiej. *Przegląd Prawno-Ekonomiczny*, 1, 97–110.

OECD, 2011, *Towards Green Growth*. Organisation for Economic Co-operation and Development, Paris. https://www.oecd.org/content/dam/oecd/en/publications/reports/2011/05/towards-green-growth_g1g1342a/9789264111318-en.pdf

Olson M., 1965, *The Logic of Collective Action: Public Goods and Theory of Groups*, Harvard University Press, Cambridge.

Olszewska J., 2013, *A Contextual analysis of the first year of a community supported agriculture pilot group in Poland. Paper presented on the 14th of June 2013 during conference: The Future of Consumerism and Well-Being in a World of Ecological Constraints*, Clark University, Worcester, MA.

Olszewska J., Trzaskowski P., 2014, *Rolnictwo Wspierane przez Społeczność partnerstwo między rolnikami a konsumentami*, IGO Instytut Globalnej Odpowiedzialności, Warszawa.

Onyszkiewicz M., Sylla M., 2025, Current Status of the Community-Supported Agriculture Model in Poland—Exploring Key Areas of Sustainable Operations. *Sustainability*, 17(7), 2.

Opitz I., Specht K., Piorr A., Siebert R., Zasada I., 2017, Effects of Consumer-Producer Interactions in Alternative Food Networks On Consumers' Learning About Food and Agriculture. *Moravian Geographical Reports*, 25(3), 181–191. https://doi.org/10.1515/mgr-2017-0016

Opitz I., Zoll F., Zasada I., Doernberg A., Siebert R., Piorr A., 2019, Consumer-Producer Interactions in Community-Supported Agriculture and Their Relevance for Economic Stability of the Farm - An Empirical Study Using an Analytic Hierarchy Process. *Journal of Rural Studies*, 68, 22–32.

Oslo Manual, 1997, *The Measurement of Scientific and Technological Activities. Proposed Guidelines for Collecting and Interpreting Technological Innovation Data*, 2nd Edition, OECD/Eurostat, Paris.

Ostrom E., 1990, *Governing the Commons: The Evolution of Institutions for Collective*, Cambridge University Press, Cambridge.

Ostrom M. R., 1997, *Toward a Community Supported Agriculture: A Case Study of Resistance and Change in the Modern Food System*, PhD dissertation, University of Wisconsin–Madison.

Ostrom M. R., 2007, Community Supported Agriculture as An Agent of Change: Is It Working?. In *Remaking the North American food system: Strategies for sustainability*, edited by Hinrichs C. C., Lyson A. T., University of Nebraska Press, Lincoln, 99–120.

Overview of Community Supported Agriculture in Europe, 2016, European CSA Research Group.

Parker J., 1974. *The Economics of Innovation: The National and Multinational Enterprise in Technological Change*. Longman, London, 13.

Parot J., Wahlen S., Schryro J., Weckenbrock P., 2024. Food Justice in Community Supported Agriculture – Differentiating Charitable and Emancipatory Social Support Actions. *Agriculture and Human Values*, 41, 685–699.

Paul M., 2018. Community-Supported Agriculture in the United States: Social, Ecological, and Economic Benefits to Farming. *Journal of Agrarian Change*, 19(1), 162–180.

Pearce D., Barbier E., Markandya A., 1990. *Sustainable Development: Economics and Environment in the Third World*. Edward Elgar Publishing Ltd., London.

Peterson H.H., Taylor M.R., Baudoin Q., 2015. Preferences of Locavores Favoring Community Supported Agriculture in the United States and France. *Ecological Economics*, 119, 64–73.

Petruzzelli M., Ihle R., Colitti S., Vittuari M., 2023. The Role of Short Food Supply Chains in Advancing the Global Agenda for Sustainable Food Systems Transitions. *Cities*, 141, 104496.

Philipon P., Chiffoleau Y., Wallet F., 2017. *Et si on mangeait local*. Éditions Quæ, Versailles, 22.

Piccoli A., Rossi A., Genova A., 2021. A Socially-Based Redesign of Sustainable Food Practices: Community Supported Agriculture in Italy. *Sustainability*, 13, 11986.

PIE, Polski Instytut Ekonomiczny, 2024. *Kapitał społeczny w Polsce*. Warszawa. Dostępne online: https://pie.net.pl/wp-content/uploads/2024/11/Raport_PIE_Kapital-spoleczny-w-Polsce.pdf

Piketty T., 2014. *Capital in the Twenty First Century*. The Belknap Press of Harvard University Press, Cambridge, London.

Pirog R., 2009. *Local Foods: Farm Fresh and Environmentally Friendly*. http://www.leopold.iastate.edu/research/marketing_files/WorldBook.pdf (accessed 2.11.2014).

Polanyi K., 1944. *The Great Transformation*. Farrar & Rinehart, New York.

Pole A., Gray S., 2013. Farming Alone? What's up with the "C" in Community Supported Agriculture. *Agriculture and Human Values*, 30, 85–100.

Ponisio L.C., M'Gonigle L.K., Mace K.C., Palomino J., de Valpine P., Kremen C., 2015. Diversification Practices Reduce Organic to Conventional Yield Gap. *Proceedings of the Royal Society B: Biological Sciences*, 282(1799), 20141396. https://doi.org/10.1098/rspb.2014.1396

Powszechny Spis Rolny 2020. *Raport z Wyników*. Główny Urząd Statystyczny, Warszawa 2021.

Prymon K., 2019. *Determinanty i kierunki rozwoju rachunkowości rolniczej*. Wydawnictwo Uniwersytetu Przyrodniczego we Wrocławiu, Wrocław.

Putnam R., 2000. *Bowling Alone: The Collapse and Revival of American Community*. Simon & Schuster, New York, London, Toronto, Sydney, Singapore.

Putnam R.D., 1993. *Making Democracy Work: Civic Traditions in Modern Italy*. Princeton University Press, Princeton.

Raftowicz M., 2020. *The climate crisis as a product of globalization. 20th International Scientific Conference Globalization and Its Socio-Economic Consequences*, 21–22.10.2020, Rajecke Teplice.

Raftowicz M., 2021. The Climate Crisis as a Product of Globalization. *SHS Web of Conferences*, 92, 06029, 1–6.

Raftowicz M., 2022. *Uwarunkowania rozwoju krótkich łańcuchów dostaw żywności*. Wydawnictwo Uniwersytetu Przyrodniczego we Wrocławiu, Wrocław.

Raftowicz M., 2023. The Importance of the Degrowth Concept in the 21st Century. *Argumenta Oeconomica*, 51(2).

Raftowicz M., Kalisiak-Mędelska M., Kurtyka-Marcak I., Struś M., 2019. *Krótkie łańcuchy dostaw na przykładzie karpia milickiego*. CeDeWu, Warszawa.

Raftowicz M., Korabiewski, B. 2025, Regional Determinants of the Development of Short Food Supply Chains in Poland. *Sustainability*, 17, 9772.

Raftowicz M., Struś M., 2026, Idea vs. Reality: Perspectives and Barriers to the Development of Community-Supported Agriculture in Poland. *Sustainability*, 18, 501.

Raftowicz-Filipkiewicz M., Le Gallic B., 2018. Perception of seafood products in Poland: An Explanatory Approach Through the Viewing of Culinary Television Programs. In *Les enjeux du développement économique, financier et écologique dans une mondialisation risquée*, edited by K. Malaga, A. Redslob, B. Woźniak-Jęchorek, Wydawnictwo Uniwersytetu Ekonomicznego w Poznaniu, Poznań.

Raftowicz-Filipkiewicz M., Struś M., Pilawka T., 2016. Rola NGO we wdrażaniu polityki spójności. Teoria a praktyka. *Prace Naukowe Uniwersytetu Ekonomicznego we Wrocławiu*, 455, 93–100.

Raworth K., 2017. *Doughnut Economics: Seven Ways to Think*. Random House Business Books, London.

Renting H., Marsden T.K., Banks J., 2003. Understanding Alternative Food Networks: Exploring the Role of Short Food Supply Chains in Rural Development. *Environment and Planning A*, 35(3), 393–411.

Renting H., Schermer M., Rossi A., 2012. Building Food Democracy: Exploring Civic Food Networks and Newly Emerging forms of Food Citizenship. *International Journal of Sociology of Agriculture and Food*, 19(3), 289–307.

Rodrik D., 2011. *The Globalization Paradox: Democracy and the Future of the World Economy*. W.W. Norton & Company, New York.

Rodrik D., 2015. *Economics rules. The rights and Wrocns of the dismal science*, W. W. Norton & Company, New York, London.

Rogala W., 1972. Działalność Państwowych Nieruchomości Ziemskich (1946–1949). *Rocznik Muzeum Rolnictwa w Szreniawie*, 5, 31.

Rogall H., 2012. *Nachhaltige Ökonomie: Ökonomische Theorie und Praxis einer nachhaltigen Entwicklung*. Metropolis, Marburg.

Rosman A., MacPherson J., Arndt M., Helming K., 2024. Perceived Resilience of Community Supported Agriculture in Germany. *Agricultural Systems*, 220, 104068.

Rossi A., Piccoli A., Feola G., 2024. Transforming Labour Around Food? The Experience of Community Supported Agriculture in Italy. *Agriculture and Human Values*, 41, 1667–1686.

Rozporządzenie w sprawie minimalnego wynagrodzenia za pracę oraz wysokości minimalnej stawki godzinowej w 2024 r. https://www.gov.pl/web/premier/min-stawka-godzinowa-2024 (dostęp 08.05.2024 r.).

Rynio D., 2013, *Kształtowanie nowej polityki regionalnej Polski w warunkach globalizacji i integracji*. Wydawnictwo Uniwersytetu Ekonomicznego we Wrocławiu, Wrocław.

Samoggia A., Perazzolo C., Kocsis P., Del Prete M., 2019. Community Supported Agriculture Farmers' Perceptions of Management Benefits and Drawbacks. *Sustainability*, 11, 3262.

Sandel M., 2012. *What Money Can't Buy: The Moral Limits of Markets*. Farrar, Straus and Giroux, New York.

Schlicht S., Volz P., Weckenbrock P., Le Gallic T., 2012. *Community Supported Agriculture: An overview of Characteristics, Diffusion and Political Interaction in France, Germany, Belgium and Switzerland*. ACTeon und Die Agronauten, Freiburg.

Schmutz U., Kneafsey M., Sarrouy Kay C., Doernberg A., Zasada I., 2018. Sustainability Impact Assessments of Different Urban Short Food Supply Chains: Examples from London, UK. *Renewable Agriculture and Food Systems*, 33, 518–529.

Schnell S.M., 2007. Food with a Farmer's Face: Community-supported Agriculture in the United States. *Geographical Review*, 97(4), 550–564.

Schumpeter J. A., 1934, *The theory of economic development: An inquiry into profits, capital, credit, interest, and the business cycle*. Harvard University Press, Cambridge, MA.

Scoones I., 2016. *The Politics of Sustainability and Resilience*. Annual Review of Environment and Resources, Vol. 41, 293-319.

Sennett R., 1998. *The Corrosion of Character: The Personal Consequences of Work in the New Capitalism*. W.W. Norton & Co, New York.

Serageldin I., Grootaert C., 2000. Defining Social Capital: An Integrating View. In *Social Capital: A Multifaceted Perspective*, edited by P. Dasgupta, I. Serageldin, World Bank, Washington, DC, 40–58.

Seyfang G., Longhurst N., 2016. What Influences the Diffusion of Grassroots Innovations for Sustainability? Investigating Community Currency Niches. *Technology Analysis & Strategic Management*, 28(1), 1–23.

Silverberg G., Soete L., 1994. *The Economics of Growth and Technical Change*. Edward Elgar, Brookfield, 21.

Six B., van Zimmeren E., Popa F., Frison C., 2015. Trust and Social Capital in the Design and Evolution of Institutions for Collective Action. *International Journal of the Commons*, 9(1), 151–176.

Skodlarski J., 2007. *Zarys historii gospodarczej Polski*. Wydawnictwo Naukowe PWN, Warszawa, 404.

Skorupska W., 2024. Kooperatywy spożywcze w Polsce jako przykład alternatywnej spółdzielczości i pozaformalnej aktywności na poziomie lokalnym w Polsce. *Studia Regionalne i Lokalne*, 4(98), 67.

Slangen L.H.G., 2001. *Sustainable Agriculture: Getting the Institutions Right*. CEESA Discussion Paper No. 1, Berlin.

Smith A., Voß J.-P., Grin J., 2010. Innovation Studies and Sustainability Transitions: The Allure of the Multi-Level Perspective and Its Challenges. *Research Policy*, 39(4), 435–448.

Smith J., Jehlička P., 2013. Quiet Sustainability: Fertile Lessons from Europe's Productive Gardeners. *Journal of Rural Studies*, 32, 148–157. https://doi.org/10.1016/j.jrurstud.2013.05.002

Social capital index, 2024. https://solability.com/the-global-sustainable-competitiveness-index/social-capital

SolAbility. https://solability.com/the-global-sustainable-competitiveness-index/social-capital

Sołdaczuk J., Misala J., 2001. *Historia handlu międzynarodowego*. PWE, Warszawa.

Śpiewak R., Błesznowski B., Gąsiorowska A., 2022. *Kooperatywy spożywcze w UE. Lekcje dla Polski*. Fundacja Kooperatywy Grochowskiej we współpracy z Fundacją im. Henricha Bölla, Warszawa.

Sproul T.W., Kropp J.D., 2015. A General Equilibrium Theory of Contracts in Community Supported Agriculture. *American Journal of Agricultural Economics*, 97, 1345–1359.

Statista, 2025. https://www.statista.com/statistics/962840/poland-monthly-minimum-wage/ and https://nbp.pl/statystyka-i-sprawozdawczosc/kursy/tabela-a

Stephens E.C., Martin G., van Wijk M., Timsina J., Snow V., 2020. Editorial: Impacts of COVID-19 on Agricultural and Food Systems Worldwide and On Progress to the Sustainable Development Goals. *Agricultural Systems*, 183, 102873.

Stiglitz J.E., Sen A., Fitoussi J.-P., 2009. *Report by the Commission on the Measurement of Economic Performance and Social Progress*. https://ec.europa.eu/eurostat/documents/8131721/8131772/Stiglitz-Sen-Fitoussi-Commission-report.pdf

Struś M., 2018, Rolnictwo Wspierane przez Społeczność – przykład innowacji konstruktywnych na obszarach wiejskich, *Problemy Drobnych Gospodarstw Rolnych* ,1, 63–71.

Struś M., Kalisiak-Mędelska M., Nadolny M., Kachniarz M., Raftowicz M., 2020. Community-Supported Agriculture as a Perspective Model for the Development of Small Agricultural Holding in the Region. *Sustainability*, 12, 2656.

Struś M., Raftowicz M., 2023. The Sustainable Development Paradigm Versus Land Concentration Processes. *Ruch Prawniczy, Ekonomiczny i Socjologiczny*, 85(3), 119–134.

Struś M., Raftowicz M., 2025. Land Concentration Processes in Poland in the Light of D.C. North's Development Paradigm. *Rozwój Regionalny i Polityka Regionalna*, 73, 47–64.

Szakál D., Balázs B., 2020. Co-creation for Transformation: Food for Degrowth in Budapest Food City Lab Initiatives. In *Food for Degrowth*, edited by G. Kallis et al., Routledge, London.

Sztompka P., 1999. *Trust: A Sociological Theory*. Cambridge University Press, Cambridge.

Szymanowski W., 2006. *Łańcuchy i sieci dostaw żywności w Polsce i w wybranych krajach europejskich – ewolucje i przykłady zastosowań technologii informacyjnych*. Dom Wydawniczy Elipsa, Warszawa.

Tanasă L., Brumă I.-S., Doboş S., 2015. The Role of Short Food Supply Chains in the Development of Small-Scale Local Producers. Case Study: Harghita County. In *Agrarian Economy and Rural Development – Realities and Perspectives for Romania*, 6th Edition of the International Symposium, November 2015, The Research Institute for Agricultural Economy and Rural Development (ICEADR), Bucharest, 286–293.

Tang H., Liu Y., Huang G., 2019. Current Status and Development Strategy for Community-Supported Agriculture (CSA) in China. *Sustainability*, 11, 3008.

The Federation of German Consumer Organisations. http://en.vzbv.de/

The Global Sustainable Competitiveness Index 2024, solability.com/news-insights/gsci-2024.

Tuck S.L., Winqvist C., Mota F., Ahnström J., Turnbull L.A., Bengtsson J., 2014. Land-use Intensity and the Effects of Organic Farming on Biodiversity: A Hierarchical Meta-Analysis. *Journal of Applied Ecology*, 51(3), 746–755. https://doi.org/10.1111/1365-2664.12219

United States Department of Agriculture, National Agricultural Statistics Service (USDA/NASS), 2014. *2012 Census of Agriculture—Summary and State Data*. USDA/NASS, Washington, DC.

URGENCI – The International Network for Community Supported Agriculture, 2021. *Enacting Resilience: The Response of Community Supported Agriculture Movements to the COVID-19 Crisis in Europe.* URGENCI, La Rochelle. https://urgenci.net/wp-content/uploads/2021/05/Enacting-Resilience-FINAL.pdf

URGENCI Country Report: Solawi cultivating solidarity in German, 2025, 5.

van der Ploeg J.D., Frouws J., 1999. On Power and Weakness, Capacity and Impotence: Rigidity and Flexibility in Food Chains. *International Planning Studies*, 4(3), 333–347.

Van En R., 1995. Eating for Your Community: Towards Agriculture Supported Community. *Context*, 42, 29–31.

van Oers L.M., Boon W.P.C., Moors E.H., 2018. The Creation of Legitimacy in Grassroots Organisations: A Study of Dutch Community-Supported Agriculture. *Environmental Innovation and Societal Transitions*, 29, 55–67. https://doi.org/10.1016/j.eist.2018.04.002

Vassalos M., Gao Z., Zhang L., 2017. Factors Affecting Current and Future CSA Participation. *Sustainability*, 9, 478.

Vávra J., Megyesi B.G., Duží B., Craig T., Klufová R., Lapka M., Cudlínová E., 2017. Food Self-Provisioning in Europe: An Exploration of Sociodemographic Factors in Five Regions. *Rural Sociology*, 82(4), 596–622. https://doi.org/10.1111/ruso.12180

Venugopal R., 2015. Neoliberalism as Concept. *Economy and Society*, 44(2), 165–187.

Vereijken P., 1997. A Methodical Way of Prototyping Integrated and Ecological Arable Farming System (I/EAFS) in Interaction with Pilot Farms. *Perspectives for Agronomy, Developments in Crop Science*, 25, 293–308.

Visser O., Dorondel S., Jehlička P., Spoor M., 2019. Post-socialist Smallholders: Silence, Resistance and Alternatives. *Canadian Journal of Development Studies*, 40(4), 499–510. https://doi.org/10.1080/02255189.2019.1688649

Volz P., Weckenbrock P., Nicolas C., Jocelyn P., Dezsény Z., 2016. *Overview of Community Supported Agriculture in Europe*. https://urgenci.net/wp-content/uploads/2016/05/Overview-of-Community-Supported-Agriculture-in-Europe.pdf

Wang H., Ma Z., Han J., 2025. Pains and Gains: Opportunities and Challenges for Grain Farmers to Participate in Short Supply Chain in China. *Agricultural and Food Economics*, 13, 53. https://doi.org/10.1186/s40100-025-00406-1

Wellner M., Theuvsen L., 2016. *Community supported agriculture als neuer Impuls für die Regionalvermarktung? Stand der Forschung und Abgrenzung von anderen alternativen Lebensmittelnetzwerken. GEWISOLA 56th Annual Conference*, Bonn, Germany.

Wellner M., Theuvsen L., 2018. Community Supported Agriculture – Determinanten der Teilnahmebereitschaft deutscher Landwirte. *GEWISOLA*.

Wezel A., Goette J., Lagneaux E., Passuello G., Reisman E., Rodier C., Turpin G., 2018. Agroecology in Europe: Research, Education, Collective Action Networks, and Alternative Food Systems. *Sustainability*, 10, 1214.

Williamson O.E., 1998. *Ekonomiczne instytucje kapitalizmu. Firmy, rynki, relacje kontraktowe*. PWN, Warszawa.

Wittenberg J., Gernert M., El Bilali H., Strassner C., 2022. Towards Sustainable Urban Food Systems: Potentials, Impacts and Challenges of Grassroots Initiatives in the Foodshed of Muenster, Germany, *Sustainability*, 14, 13595.

Witzling L., Shaw B., Strader H.C., Sedlak C., Jones E., 2020. The Role of Community: CSA Member Retention. *British Food Journal*, 122(7), 2289–2302.

Woods T., Ernst M., Tropp D., 2017. *Community Supported Agriculture – New Models for Changing Markets*. U.S. Department of Agriculture, Agricultural Marketing Service, Washington.

Woods T., Tropp D., 2015. CSAs and the Battle for the Local Food Dollar. *Journal of Food Distribution Research*, 46(2), 17–29.

Woś A., Zegar J. S., 2002. *Rolnictwo społecznie zrównoważone*, IERiGŻ-PIB, Warszawa.

Zajda K., Mazurek D., 2022. Public Institutions and NGOs Cooperation for Social Innovations in Post-Socialist Rural Poland. *European Countryside*, 14(4), 623–637.

Zbyrowska M.J., 2022. Zasady z Rochdale w obecnej działalności polskich kooperatyw spożywczych. *Metody Ilościowe w Badaniach Ekonomicznych*, 23(1), 24–33.

Zegar J. S., 2009, Kwestia koncentracji ziemi w polskim rolnictwie indywidualnym, *Roczniki Nauk Rolniczych, Seria G*, 96, (4).

Zegar J.S., 2012. *Współczesne wyzwania rolnictwa*. PWN, Warszawa.

Zepeda L., Reznickova A., Russell W.S., 2013a. CSA Membership and Psychological Needs Fulfillment: An Application of Self-Determination Theory. *Agriculture and Human Values*, 30(4), 605–614.

Zepeda L., Sirieix L., Pizarro A., Coderre F., Rodier F., 2013b. A conceptual framework for analyzing consumers' food label preferences: an exploratory study of sustainability labels in France, Quebec, Spain and the US. *Post-Print hal-02650512*, HAL.

Zoll F., Kirby C.K., Specht K., Siebert R., 2023. Exploring Member Trust in German Community-Supported Agriculture: A Multiple Regression Analysis. *Agriculture and Human Values*, 40(2), 709–724.

Zukunftskommission Landwirtschaft (ZLK), 2021. *Zukunft Landwirtschaft. Eine gesamtgesellschaftliche Aufgabe. Empfehlungen der Zukunftskommission Landwirtschaft*.

Index

For Product Safety Concerns and Information please contact our EU representative GPSR@taylorandfrancis.com Taylor & Francis Verlag GmbH, Kaufingerstraße 24, 80331 München, Germany

Printed and bound by CPI Group (UK) Ltd, Croydon, CR0 4YY
14/09/2026
02216243-0008